T0295065

Inventory Planning
with Innovation

Inventory Planning with Innovation

A Cost Focus

by

Sanjay Sharma

CRC Press
Taylor & Francis Group
Boca Raton London New York

CRC Press is an imprint of the
Taylor & Francis Group, an **informa** business

First edition published 2021
by CRC Press
6000 Broken Sound Parkway NW, Suite 300, Boca Raton, FL 33487-2742

and by CRC Press
2 Park Square, Milton Park, Abingdon, Oxon, OX14 4RN

Library of Congress Cataloging-in-Publication Data

Names: Sharma, Sanjay (Professor of industrial engineering), author.
Title: Inventory planning with innovation : a cost focus / Sanjay Sharma.
Description: Boca Raton : CRC Press, 2021. | Includes bibliographical references and index. | Summary: "Discussing innovation efforts and different factors including the type and size of an organization, nature, and boundary of innovation, and area and scope of work, this text will be a valuable resource for senior undergraduate, graduate students, and professionals in the field of industrial engineering, production engineering, and manufacturing science. It further covers important concepts such as related expenditure, time consideration, and related total cost planning in a comprehensive manner"– Provided by publisher.
Identifiers: LCCN 2020049007 (print) | LCCN 2020049008 (ebook) | ISBN 9780367740986 (hardback) | ISBN 9781003156093 (ebook)
Subjects: LCSH: Inventory control.
Classification: LCC TS160 .S5135 2021 (print) | LCC TS160 (ebook) | DDC 658.7/87–dc23
LC record available at https://lccn.loc.gov/2020049007
LC ebook record available at https://lccn.loc.gov/2020049008

ISBN: 978-0-367-74098-6 (hbk)
ISBN: 978-1-003-15609-3 (ebk)

Typeset in Times
by Deanta Global Publishing Services, Chennai, India

Contents

Preface

Product innovation is necessary in order to become competitive in certain industrial/ business areas. However, in many cases, the product life cycle is relatively short. Even in the event of uniform demand for a certain period, this period is relatively short. Traditional inventory planning is usually for a longer period, along with the estimation of annual total cost and its optimization. When innovation or frequent innovation is a critical component for any industry/business, a suitable strategically decided period is a prime requirement for subsequent inventory planning with the focus on cost. Depending on the specific scenario, this strategic period might be of two to three years or even a few months. This book is concerned with the inventory planning for such a stated period, along with the innovation and an associated total cost focus.

This book is expected to be a useful reference source for both undergraduate and postgraduate students in engineering and management. It has good potential for elective/supplementary core courses.

Author Biography

Dr. Sanjay Sharma is Professor at National Institute of Industrial Engineering (NITIE), Mumbai. He is an operations and supply chain management educator and researcher. He has more than three decades of experience in various fields: industrial, managerial, teaching/training, consultancy, and research. He has to his credit many awards and honors. He has published eight books and research papers in various journals such as *European Journal of Operational Research, International Journal of Production Economics, Computers & Operations Research, International Journal of Advanced Manufacturing Technology, Journal of the Operational Research Society,* and *Computers and Industrial Engineering.* He is a reviewer for several international journals. He is also on the editorial board of a few journals, including *International Journal of Logistics Management.*

1 Introduction

In the case of inventory planning, it becomes necessary to differentiate between procurement and production activities. In order to link innovation efforts with inventory issues, such efforts are first discussed. Also, the factors influencing innovation efforts are explained in sufficient detail. These include the type and size of an organization, nature and boundary of innovation, and area and scope of work. Additionally, related expenditure and time considerations are suitably incorporated. Strategic period is also introduced when an innovation and the related efforts are explicitly included in the analysis.

1.1   INNOVATION EFFORTS

An organization requires innovation efforts in terms of products and processes. This requirement might arise because of several factors such as business competition, survival, and enhancement of benefits including profit. Business leaders/promoters often believe that a new way of thinking for product designs/processes helps with market growth, eventually leading to the generation of benefits. The interdependence of existing products/processes, innovation activities, and appropriate benefits is shown in Figure 1.1.

The current situation of a business should be well understood in order to initiate innovation efforts and also to ensure eventual success. The amount of effort put into innovation depends on the existing situation. Innovation success lies in the benefits generated and depends on innovation efforts and also on their implementation in the existing industrial/business scenario.

The level of efforts needed in the context of innovation depends on many factors, as shown in Figure 1.2.

1.1.1   TYPE OF ORGANIZATION

The type of organization influences innovation-related activities. It may include manufacturing as well as service sectors. The manufacturing sector covers a wide variety of products, whereas the service sector includes education, health, and a wide variety of corporate services. As per the type of organization, the appropriate resources should be made available for starting innovation-related activities. It needs to be ascertained whether the desired expertise/resources are available within the organization or whether they need to be arranged from outside. This is important in order to estimate the investment or expenditure required for the completion of intended activities for improvement in either products or processes associated with the concerned type of organization.

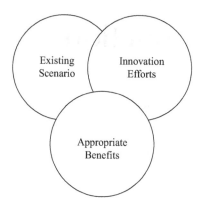

FIGURE 1.1   Relevant interaction.

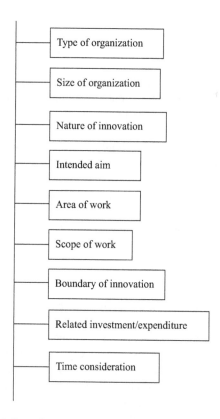

FIGURE 1.2   Factors influencing the innovation efforts.

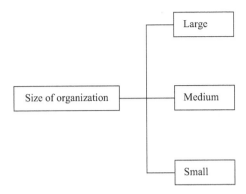

FIGURE 1.3 Relative size.

1.1.2 SIZE OF ORGANIZATION

Figure 1.3 shows the categorization pertaining to the relative size of the organization. Furthermore, the perception of relative size might depend on the following factors:

(a) Country
(b) Consumption
(c) Fixed asset
(d) Resources available
(e) Investment

In a large-sized organization, planning and approval for the start of innovation-related activities may take considerable time. However, a large investment can be decided depending on the need. Additional resources such as professionals with relevant expertise might also be available within the organization. On the other hand, in a smaller organization, decisions related to approval for an innovation activity might be taken quickly. However, there can be investment constraints, and also human resources with required expertise may not be available within the enterprise. Medium-sized organizations can be at an intermediate level with regard to these issues.

1.1.3 NATURE OF INNOVATION

This aspect may be concerned with the improvement level as a potential result of an innovation activity: that is, whether the potential improvement is little or great; whether a radical improvement is the end result or whether the activity may end with certain stepwise modifications in a particular area of work.

1.1.4 INTENDED AIM

The main aim for starting an innovation effort should be ascertained explicitly. For example, the aim can be cost reduction. This cost reduction might be concerned with a product or a process. However, any predetermined aim can be accomplished by activities such as:

   (i)   Product design modification
   (ii) Substitution of material
   (iii) Process design modification
   (iv) Improvement in the existing production process

Another aim can be associated with the expectation of a higher price for the organization's product or services. This aim can be accomplished by activities such as:

   (a) New product development for select customer segment
   (b) Value creation in the provided services

Appropriate steps in order to achieve the intended aim may then be taken or planned in a phased manner.

1.1.5   AREA OF WORK

A manufacturing organization has many divisions related to various areas of work:

   (i)   Purchase/procurement
   (ii) Production/manufacturing
   (iii) Inspection/testing/quality control

There are several other areas as well. Similarly, in other kinds of organizations also, there are different spheres of work. It should be ascertained what is the focus area at the start of the related innovation activities.

1.1.6   SCOPE OF WORK

As shown in Figure 1.4, the scope of work may encompass planning aspects, or it may cover technological aspects also.

   For example, certain innovation consideration is included in inventory planning of an organization. Therefore, the core technology is broadly outside the scope of work. However, when certain engineering innovation is the objective in a modification of the existing production process, then technology is primarily within the scope.

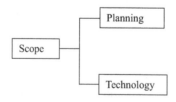

FIGURE 1.4   Relevant scope.

1.1.7 BOUNDARY OF INNOVATION

In some cases, an innovation activity may only impact the organization from within. For example, consider a scenario when a modification in a component produced within the factory premises itself is planned. In such a case, there may not be a significant need to involve outside parties. But when an assembly maker wishes to modify the final product, there might be a need to make changes in a particular component design. In such a situation when an outside party supplies the component, there is a need to involve the party at a suitable time. A boundary of the innovation may thus be conceptualized as follows:

(a) Within the organization
(b) Beyond the organization

1.1.8 RELATED INVESTMENT/EXPENDITURE

The level of investment or expenditure should be estimated before starting innovation efforts. The level of investment, whether high or low, will depend on the complexity of the innovation process. Figure 1.5 shows various aspects pertaining to this complexity.

In certain cases, a piece of specialized equipment may be needed. This might relate to:

(i) Manufacture
(ii) Inspection
(iii) Testing

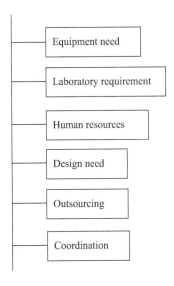

FIGURE 1.5 Various aspects related to innovation process.

Additionally, there might be laboratory requirements. It needs to be ascertained whether specifically qualified and experienced human resources are available. In the context of a manufacturing organization, clarify whether component design is part of the whole innovation process, either directly or indirectly. Furthermore, scope always exists for outsourcing some of the activities. A study can be made related to coordination efforts, particularly concerning outsourced activities. Such considerations help a lot in a precise estimation of the related expenditure or investment need attributing to the overall innovation efforts.

1.1.9 TIME CONSIDERATION

After an estimation of the related expenditure pertaining to the innovation efforts, it should be established when such expenditure will occur. That is, the time consideration is associated with the investment/expenditure attributed to the innovation efforts.

For example, ₹1 million is the total expenditure. Now the possibilities are as follows:

(a) The entire expenditure occurs at the beginning of the planned period.
(b) The whole expenditure happens throughout the planned period in a phased manner, i.e., in subperiods.

For instance, if the planned period is one year, then consider the first possibility with the subperiod as three months or a quarter of the year. Suppose that the opportunity cost (or also the cost of capital) is 10% per year. In such a case, the total expenditure might be assumed as ₹1.1 million.

For the quarter as subperiod, an apportioned innovation cost can be estimated as follows:

$$\frac{1.1}{4} = ₹0.275 \text{ million}$$
$$= ₹275,000$$

Similarly, for the second possibility, a suitable estimation for the apportioned innovation cost (pertaining to the innovation efforts) is made. However, when the subperiod is not relatively long (in general) and also depends on the perception of management toward the expenditure, time consideration might be ignored. In such a case, i.e., when the total cost is incurred in a phased manner, an apportioned innovation cost for the quarter as subperiod might be:

$$\frac{1}{4} = ₹0.25 \text{ million}$$
$$= ₹250,000$$

Table 1.1 represents the period and subperiod.

TABLE 1.1
Period and the Corresponding Subperiod

Planning Period	Subperiod
Decade	Year
Half decade	Year
Year	Quarter or month
Quarter	Month
Month	Week (or day)

Therefore, it is possible to estimate the apportioned innovation cost (associated with innovation efforts) depending on:

(i) Period
(ii) Subperiod
(iii) Perception of management toward the time consideration for investment/expenditure concerning innovation efforts

1.2 INVENTORY PLANNING

Usually a cyclic approach is implemented for many items in an organization in order to plan for inventories, either directly or indirectly. Inventories are primarily categorized into:

(a) Purchase inventory
(b) Manufacturing inventory

1.2.1 PURCHASE INVENTORY

Purchase or procurement inventory is concerned primarily with input items. These input items are procured from outside sources generally. As shown in Figure 1.6, purchase inventory can also be related to different consumption segments.

In addition to raw materials and purchased components, spare parts are also procured and consumed regularly. In any factory, fuel and welding electrodes among many other items are needed in various processes. Office items such as stationery and printer cartridges also constitute an important segment for consumption. Other items might include products such as furniture.

For most of the materials, a cyclic approach is usually implemented in order to plan for inventory items. This cyclic approach consists of a planning period divided into an appropriate subperiod. It is also shown in Figure 1.7.

1.2.2 MANUFACTURING INVENTORY

In a manufacturing company, raw material or input item is fed to various facilities arranged in a certain layout. An input item is converted into a suitable component

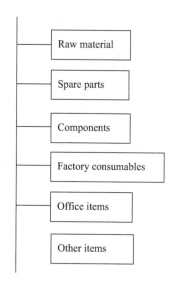

FIGURE 1.6   Consumption segments for purchase inventory.

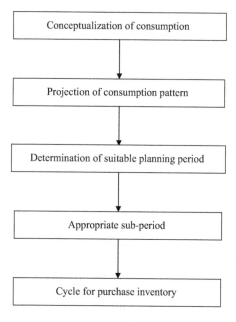

FIGURE 1.7   Approaching purchase inventory cycle.

for further processing. A simplified scheme is shown in Figure 1.8 for the production process. Facilities are arranged and set up in a layout depending on the product being manufactured. In a particular pattern, the work-in-progress might move from one facility to the next. In this way, a smooth material flow can be made possible from the predecessor to a successor facility. A component might move throughout the facility layout, and at the end of the journey, a finished item can be made

Finished item

Facility/
Facilities

Input item

FIGURE 1.8   Simplified scheme for production process.

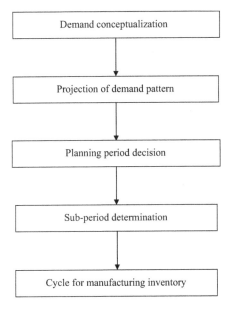

Demand conceptualization

Projection of demand pattern

Planning period decision

Sub-period determination

Cycle for manufacturing inventory

FIGURE 1.9   Approach to manufacturing inventory cycle.

available for onward transportation/storage as per the business/customer need precisely. Manufacturing inventory may relate to a component/finished item in terms of production inventory/produced inventory.

For many components/finished items, a cyclic approach is usually implemented directly or indirectly. An approach to the manufacturing inventory cycle is shown in Figure 1.9. A demand is conceptualized and projected in terms of the demand pattern. The cycle is approached after deciding a planned period and also a subperiod.

1.3   STRATEGIC PERIOD

A planned duration should be decided for inventory planning. For innovative items, such a strategic period may depend on a few aspects, as shown in Table 1.2.

TABLE 1.2

Certain Aspects Pertaining to a Strategic Period

Few Aspects	Significance in Terms of Strategic Period
Type of industry	Innovation significance
Nature of product	Lead time significance
Projected demand	Continuance
Nature of investment	Duration
Area of work	Functional differentiation

An appropriate strategically planned period may depend on the type of industry, and the degree of significance of the innovation-linked activities in that kind of industry. There is a certain lead time for a particular product, i.e., the difference in time when finally a product is made available after getting an order for it from the customer. Depending on the nature of the product, this lead time is lower or higher. This aspect affects the strategic period for inventory planning along with the innovation efforts. Demand for an item can be projected for a few years or a few months also among other time zones. For instance, if demand is uniform for an innovative item currently, then an important question is:

How long will it continue in a similar pattern?

The nature of investment is also linked to its relevant duration. That is, whether the expenditure is made at the beginning, and the length of the period the inventory plan would be appropriate. Alternatively, the investment is made in a periodic or phased manner. And the qualitative or quantitative effects can be assessed on the length of the strategic period.

Functional differentiation is necessary in the context of the area of work, for example, whether the area of work is purchase/procurement or production/manufacturing. Depending on the complexity of environment, a strategic period may differ for the organization in the context of a functional area.

After the conceptualization of a strategic period, a planned duration can be a key parameter for:

(i) Purchase function
(ii) Manufacturing function

1.3.1 Planned Duration for Purchase

There can be a wide variety of procurement or purchase situations being faced in commercial or economic environment. Multiple tiers can also be found in practice for procurement of items and also selling these finally to intended consumers or a group of such consumers. In order to develop or introduce this concept, one simple scenario for purchase function can be considered that relates to a trading house. In a trading firm, an innovative product might be purchased in certain quantities from a suitable producer among others. These products are available for sale to individual customers as shown in Figure 1.10.

Along with the regular items, such an innovative product is also purchased and stocked by the trading firm. Because of the nature of the innovative item and the associated uncertainty, the product life cycle might be shorter. Therefore the total cost estimation should be made available for a suitable planned duration in such cases, and it should be optimized for this duration completely along with the innovation efforts. Such efforts may relate to the development of a certain level of understanding of this item. It also requires certain time and effort in order to sell this type of product while generating awareness among customers toward the item.

In order to estimate the total cost for a planned duration, it is necessary to choose this duration wisely, besides including multiple qualitative and quantitative aspects. There are also occasions when such duration can excessively be shorter or longer. If the total planned duration is relatively shorter, there is a possibility of loss in terms of economies of scale for a few factors, as shown in Figure 1.11, either directly or indirectly. This happens because such activities get repeated frequently. Each time an order is placed for such items, it needs arrangements for transportation. This results in ordering efforts several times, and also transporting a relatively small number of items frequently. When the ordered items reach the trading house, unloading and also receiving of the lot has to be exercised. Similarly, the paperwork on receiving the lot and also the unloading efforts are repeated multiple times in the chosen time period. Wherever relevant, inspection, including visual inspection, may also be carried out for a certain time. In many cases, a portion of the inspection effort is attributed to the whole lot of items and this might be unnecessarily more frequent. Thus, a loss in terms of the economies of scale may occur.

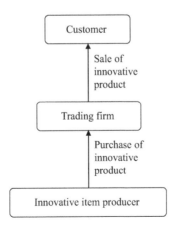

FIGURE 1.10   Innovative item procurement.

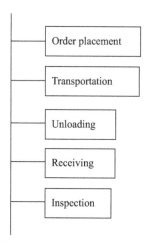

FIGURE 1.11   Factors concerned with a possible loss in terms of economies of scale.

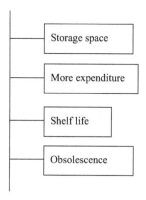

FIGURE 1.12   Factors associated with overstocking.

On the other hand, the total planned duration may be relatively longer. In order to take advantage of economies of scale, there may be overstocking of items. This is potentially on account of economy:

(i) Order placement
(ii) Transportation
(iii) Unloading
(iv) Receiving and inspection

This situation of overstocking of items may also result in associated factors, as shown in Figure 1.12. Because of overstocking of items, storage space can be a matter of concern. This is because a constraint on storage space is found in many successful trading organizations where already several regular products are kept for downstream consumption periodically. While overstocking, purchase costs among others are also incurred. Therefore, it can lead to a relatively greater expenditure or investment at times. In cases where the item has a certain shelf life, it might expire before serving its purpose. Expiry of the product might happen even before a customer appears in the trading firm or a customer order is received by such a firm. A particular item may become obsolete in the market due to severe competition from other products, i.e., the availability of better items meanwhile. The longer it takes to completely offload an overstock, the greater the chance that a better or more competitive product in terms of other aspects might become available for potential consumers.

Therefore, the total planned duration should be selected judiciously as well as strategically, considering all possible factors qualitatively and quantitatively.

1.3.2 PLANNED DURATION FOR MANUFACTURE

When an innovative item is to be manufactured, a suitable demand level may be projected for a certain duration, if it is possible. This helps in the total demand estimation for the planned duration. Additionally, innovation efforts also need to be visualized in order to arrive at the corresponding cost or expenditure. These efforts may include the research and development activity, a complete design of the item, and coordination activities for successful completion of the intended objective.

As such items are not regular products and also because of ongoing competition, a planned duration related to a certain strategic period might be relatively shorter. However, the total cost should be assessed and optimized for this complete duration in order to have a wider perspective. This also helps indirectly in determining an appropriate price for such innovative products, as the related efforts and the manufacture for the total duration should be suitably rewarded in the business.

The total planned period should be carefully chosen as it need not be too short or too long. If it is too long, then the foreseen scenario, including the sales pattern among other aspects, may become erroneous. This can result in overproduction. The factors associated with this are represented by Figure 1.13. A sales pattern mismatch may happen if the actual sales are lower than the expected ones. Excess stock is an outcome of this. However, stage-wise stock may be observed in order to go into the details of the problem. Certain stock might be at an intermediate stage as well as at the final stage. This results in storage space issues at various places:

(a) Intermediate space between facilities
(b) Storage area at the last stage of manufacture
(c) Warehousing area

On the other hand, the potential benefits of economies of scale may not be availed if the total planned period is too short. This is because a certain set of activities which could have been performed less frequently are done frequently. Figure 1.14 shows the activities linked to diseconomies of scale either directly or indirectly.

Each time a machine is set up, there is an internal material-handling effort. Similarly, a frequent loading of items may happen along with the shipping or transportation of relatively small quantities of a particular product.

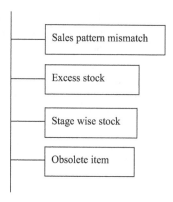

FIGURE 1.13   Factors concerned with overproduction.

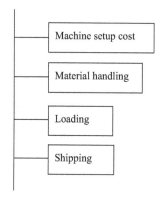

FIGURE 1.14   Activities linked to potential diseconomies of scale.

The total planned period should thus be carefully chosen in order to avail the potential benefits, particularly in the context of an innovative product.

Innovation efforts have already been introduced and explained in this chapter. The aim is to link innovation efforts with inventory planning. Various factors influencing the innovation efforts are described in sufficient detail. This also helps in understanding a certain level of investment or expenditure that might be estimated before starting the innovation efforts. Inventory planning has been categorized into two types: the purchase inventory and the manufacturing inventory. These environments are discussed for a better conceptual learning of the cyclic approach in the context of a planned strategic period.

Chapter 2 deals with the procurement inventory where purchase of an item or a component may happen in a trading organization as well as in a manufacturing organization. The role of innovation needs to be understood in an overall context of profitability and in the specific context of conversion of innovation efforts to the related cost. Therefore, this role is explained in necessary detail in order to utilize it for inventory planning. Total cost estimation is done for the planned duration as a strategic period. The total planned duration is divided into a suitable cyclic approach for the procurement inventory plan. Several procurement examples are also provided along with rigorous numerical and analytical details. This includes implications of specific ratios in the context of procurement inventory analysis along with an innovation cost.

Production inventory is covered in Chapter 3. This includes innovation efforts and related cost in a manufacturing setup. Total cost planning incorporates the holding cost, production facility setup cost, and the apportioned innovation cost. Production cycle and its continuation is determined on the basis of total cost optimization. Influence of the innovation cost and its upward and downward variation are studied. Significance of restoring the production cycle time is highlighted, and the remedial measure is incorporated in the event of such issues in the organization. Variation in the planning horizon is explained along with the associated effects. An imperfect manufacturing batch may be a reality in practice, and thus it is also analyzed for the concerned inventory planning along with the innovation.

The need for and the significance of multiple products are established in Chapter 4. The contributing factors are explained, such as consumption pattern, storage and production capacity, and the product family. Thus, the significance of multiple items is discussed in the context of procurement as well as production situation. Such problems are analyzed because there are cases where multiple items are procured as well as manufactured in a suitable time. Innovative product entry as well as exit is especially incorporated into the multi-item scenario. For innovative items, total planned duration in the context of a strategic period should be carefully selected. Examples pertaining to such strategic duration are provided.

Additional benefits of the proposed approach are given in Chapter 5. Finally, this chapter also makes available the future scope and concluding remarks. Such remarks are concerned with a single-item and multi-item purchase. These are also associated with single-item as well as multi-item production. With a wide scope in terms of application, a discussion is made in the context of organization, i.e., within an organization and beyond. An overall benefit lies in a fresh approach toward linking the inventory analysis with an innovation effort and the concerned cost.

2 Procurement Inventory

Procurement inventory basically relates to the purchase situation in a part of the business/industry. In order to estimate the total cost while including the innovation efforts explicitly, it is necessary to understand the role of innovation and also the purchase context.

2.1 PURCHASE

As shown in Figure 2.1, a purchase might be for trading as well as for manufacture.

2.1.1 PURCHASE FOR TRADING

An established trading firm may purchase regularly certain items and sell these to consumers/group of consumers. In most cases, the item is well known to the trading firm as well as to the consumers. Specifications of the product including advantages and relative disadvantages might be clearly known to all the concerned parties/people. Familiarity with such regular items may also extend toward price, volume, weight, and availability as per the relevance of such characteristic features. Overall demand might also be more or less uniform, or fluctuation with respect to certain time zones can be predicted with relative ease.

Figure 2.2 provides suitable background for cyclic purchase of regular products by the trading firm. Existing purchase pattern of the trading firm should be observed to link it to the buying pattern of consumers/group of consumers. For most of the regular items, a cyclic approach for purchase may be adopted finally in order to streamline the process. This is also associated with the balance of conflicting cost components.

2.1.2 PURCHASE FOR MANUFACTURE

A component can be purchased by a certain production firm for further processing, so that a final item is made available. For convenience, consider that a final item is composed of two components, as shown in Figure 2.3.

Of the two components, one component might be produced within the manufacturing firm from the raw material, whereas the other component may be purchased from outside.

Generally speaking, the purchased component requirement needs to be generated. The following factors may help in the generation of such requirement:

(i) Consumption pattern of the final item
(ii) Price of the component
(iii) Quality and availability of the component
(iv) Additional stock to be kept for this component

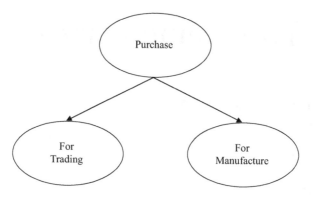

FIGURE 2.1   Purchase types.

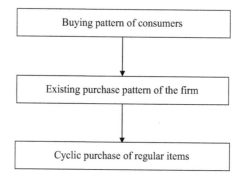

FIGURE 2.2   Background for cyclic purchase.

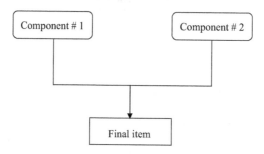

FIGURE 2.3   Two components in a final item.

For a regularly purchased component, there is a possibility for standardization in procurement process along with the characteristic features of such a purchased item. In case of certain fluctuation, including that in the requirement as well, it may be predicted with relative ease. Because of the convenience in implementation, a cyclic approach may be adopted for many purchase examples concerning further manufacturing objectives as shown in Figure 2.4.

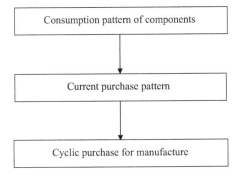

FIGURE 2.4 Approach to cyclic purchase for manufacture.

The current purchase pattern should be closely observed to link it to the consumption pattern of an input item. In order to improve the existing process, a modified approach for cyclic purchase may be explored. This approach also includes the balancing of different operational factors.

2.2 ROLE OF INNOVATION

The role of innovation needs to be understood in an overall context of profitability and in the specific context of conversion of innovation efforts to the related cost. For a regular item, a trading firm representative or seller needs minimal interaction with the customer. In many cases, the customer directly demands a regular or standard product with which he is very much familiar. The seller has negligible interaction with the buyer and a transaction may be over in a very few minutes. Traditionally, the profit may simply be evaluated by deducting the purchase cost from the sales price.

However, the buyer might not be familiar with the innovative product in another situation, and the transaction may consume several minutes. There is significant interaction between the seller and a potential buyer. Depending on the product, this major interaction may consist of the following aspects:

(a) Specifications of the product
(b) Operational features
(c) Different settings
(d) Price differentiation corresponding to accessories

This differentiation between regular and innovative products is also shown in Figure 2.5.

Another issue is that all the potential buyers do not finally become the real buyer with an ultimate billing. Because of these problems, the profitability might also get affected by an innovative item in many cases. The real costs are more than the traditional purchase cost, and thus the profit margin might be on the lower side.

Therefore, the apportioned innovation cost should explicitly be captured along with other cost components. After an analysis of the innovation efforts, it needs to

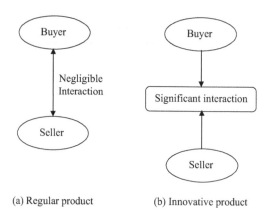

(a) Regular product   (b) Innovative product

FIGURE 2.5   Interaction level for regular/innovative product.

be converted to the cost. Finally, this should be linked to the quantities in the cyclic approach for an overall planned period.

Similarly, the role of innovation efforts can be visualized for the purchase of a certain input item for further processing at the manufacturing firm location. In case where this is a standard item being procured regularly, less time and effort are consumed. However, when this is an innovative item, the purchasing firm has to devote considerable time and resources. Such time and efforts including coordination activity relate to a wide discussion concerned with the precise requirement of the buyer. An apportioned innovation cost estimation, depending on the mentioned efforts, should also more or less take into consideration other aspects such as total expected demand, strategic period for the planned duration, and cyclically purchased number of items. In the total cost estimation, this apportioned innovation cost has a role in formulation and subsequent analysis.

2.3   TOTAL COST ESTIMATION

Consider an example where 900 units of demand are estimated over a period of 3 years, or 36 months, with almost uniform consumption throughout. In a total period of 36 months, 900 units are to be purchased for an innovative item. Therefore,

The consumption per month $= \dfrac{900}{36}$

In case where a procurement cycle is of 4 months,

the procurement quantity per cycle $= \dfrac{900}{36} \times 4$

In order to generalize, refer to Figure 2.6 where

t = cycle time for procurement in months

T = Total planned duration in months

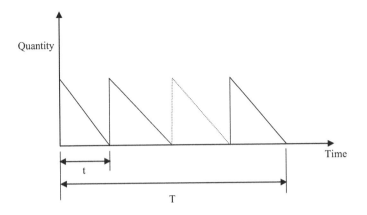

FIGURE 2.6   Procurement pattern over a total planned duration.

Now,

Procurement quantity per cycle $= \dfrac{Dt}{T}$,

where D = total demand for the planned duration (T)

Average quantity during the cycle $= \dfrac{Dt}{2T}$

Inventory cost during the cycle $= \dfrac{Dt}{2T} \cdot H \cdot t$

where H = Inventory carrying cost per unit for one month

Total inventory cost for the planned duration (T) $= \dfrac{Dt}{2T} \cdot Ht \cdot \dfrac{T}{t}$

$$= \frac{DtH}{2} \qquad\qquad (2.1)$$

$$\text{Total ordering cost} = \frac{T}{t}C \qquad\qquad (2.2)$$

where C = fixed ordering cost

$$\text{Total innovation cost for procurement} = \frac{T}{t}I \qquad\qquad (2.3)$$

where I = Apportioned innovation cost for one cycle

Estimated total cost (E) can be obtained by adding Eqs. (2.1), (2.2), and (2.3) as follows:

$$E = \frac{DtH}{2} + \frac{T}{t}C + I) \qquad\qquad (2.4)$$

Optimal value of t can be obtained by differentiating Eq. (2.4) with respect to t and equating to zero:

$$\frac{dE}{dt} = \frac{DH}{2} - \frac{T(C+I)}{t^2} = 0$$

$$\text{Or } \frac{T(C+I)}{t^2} = \frac{DH}{2}$$

$$\text{Or } t^2 = \frac{2T(C+I)}{DH}$$

$$\text{Or } t = \sqrt{\frac{2T(C+I)}{DH}} \qquad (2.5)$$

2.4 PROCUREMENT EXAMPLES

Cyclic procurement deals with purchase of a certain number of items in an appropriate cycle time. A basic example is discussed next.

2.4.1 BASIC EXAMPLE

In view of multiple factors, a strategic period in terms of a total planned duration is decided by the firm as 36 months. In this duration, a total of 900 innovative components are expected to be purchased with more or less uniform consumption. Inventory holding cost per unit component is estimated to be ₹2 per month. Fixed ordering cost per procurement cycle is ₹200, and an apportioned innovation cost per cycle is also assessed at ₹200. Now, in order to find out the cycle time for procurement in months, Eq. (2.5) can be used:

$$t = \sqrt{\frac{2T(C+I)}{DH}}$$

As $T = 36$ months

$D = 900$ number of components

$H = ₹2$

$C = ₹200$

$I = ₹200$

The procurement cycle time can be obtained as

$t = 4$ months

A significance of $\left(\dfrac{D}{T}\right)$ ratio is explained in the next example.

2.4.2 Significance of $\left(\dfrac{D}{T}\right)$ Ratio

In the basic illustration,
   $D = 900$ and $T = 36$ months
   Therefore,

$$\frac{D}{T} = \frac{900}{36} = 25$$

The ratio $\left(\dfrac{D}{T}\right)$ may increase or decrease in a business. Thus it is of interest to know the effect on cycle time for procurement.
   For instance, consider

$$\frac{D}{T} = 30$$

and the remaining parameters from the basic example as

   $H = ₹2$
   $C = ₹200$
   $I = ₹200$

Now, from Eq. (2.5),

$$t = \sqrt{\frac{2(C+I)}{H(D/T)}}$$

$$\text{Or } t = \sqrt{\frac{2 \times 400}{2 \times 30}}$$

$$= 3.65 \text{ months}$$

Table 2.1 shows the cycle time in months (t) corresponding to different $\left(\dfrac{D}{T}\right)$ ratios.
   Procurement cycle time decreases with the increase in $\left(\dfrac{D}{T}\right)$. However, a proportional increase in $\left(\dfrac{D}{T}\right)$ may be a suitable factor for a generalized approach in order to know the variation in t.
   Let
   $L = \%$ increase in $\left(\dfrac{D}{T}\right)$
   $t_m = $ value of t with $\%$ variation in $\left(\dfrac{D}{T}\right)$
   Now,
   $\%$ decrease in $t = 100\left(\dfrac{t-t_m}{t}\right) = 100\left(1-\dfrac{t_m}{t}\right)$

TABLE 2.1

Corresponding Procurement

Cycle Time for $\left(\dfrac{D}{T}\right)$ Ratios

S. No.	$\left(\dfrac{D}{T}\right)$	t
1	25	4.00
2	30	3.65
3	35	3.38
4	40	3.16
5	45	2.98

where $t_m = \sqrt{\dfrac{2(C+I)}{H(D/T)(1+L/100)}}$

$$t = \sqrt{\dfrac{2(C+I)}{H(D/T)}}$$

Therefore, the % decrease in $t = 100\left[1 - \sqrt{\dfrac{1}{(1+L/100)}}\right]$

Table 2.2 shows the percentage reduction in t.

Similarly, when the $\left(\dfrac{D}{T}\right)$ ratio reduces, the procurement cycle time increases. As the proportional reduction in the $\left(\dfrac{D}{T}\right)$ ratio can be a better factor for comparison, let

L = % reduction in $\left(\dfrac{D}{T}\right)$

t_m = value of t with % variation in $\left(\dfrac{D}{T}\right)$

TABLE 2.2

Percentage Reduction in t

S. No.	L	% reduction in t
1	3	1.47
2	6	2.87
3	9	4.22
4	12	5.51
5	15	6.75

Now, % increase in $t = 100\left(\dfrac{t_m - t}{t}\right)$

$$= 100\left(\dfrac{t_m}{t} - 1\right)$$

where $t_m = \sqrt{\dfrac{2(C+I)}{H(D/T)(1-L/100)}}$

$$t = \sqrt{\dfrac{2(C+I)}{H(D/T)}}$$

Therefore, the % increase in $t = 100\left[\sqrt{\dfrac{1}{(1-L/100)}} - 1\right]$

Table 2.3 shows the percentage increase in t. In comparison with the previous scenario, the % variation in t is higher.

2.4.3 IMPLICATIONS OF $\left(\dfrac{D}{T}\right)$ INCREASE

If the ratio $\left(\dfrac{D}{T}\right)$ increases, t value reduces as explained before. When the cycle time for procurement gets reduced, then the implications are not confined to only such a buying firm. This reduction in t also needs to be communicated to the supplying firm concerned with an innovative component, as shown in Figure 2.7.

The related factors which explicitly or implicitly get affected may include in the context of buying firm:

(i) Ordering process
(ii) Purchase lead time
(iii) Storage space

TABLE 2.3
Percentage Increase in t

S. No.	L	% increase in t
1	3	1.53
2	6	3.14
3	9	4.83
4	12	6.60
5	15	8.47

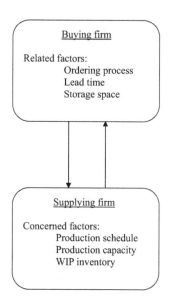

FIGURE 2.7   Communication and implications for the firms.

There is also a two-way communication between the buying and supplying firms regarding a reduction in the procurement cycle time at the buying firm side. The supplying firm may or may not agree to this change. This is because the concerned factors at the supplying firm operations get affected either explicitly or implicitly. Such factors may include:

(a) Production schedule
(b) Production capacity
(c) Work-in-process inventory

In order to avoid uncertainty concerned with an acceptance of this change by the supplying firm particularly with respect to the time aspects and also coordination/communication efforts, the buying firm may prefer to explore internally.

While doing so, an important factor is the inventory holding cost per unit item per month. With reference to the previous basic example:

$T = 36$ months
$D = 900$
$C = ₹200$
$I = ₹200$
$H = ₹2$
$t = 4$ months

With an increase in $\left(\dfrac{D}{T}\right)$, i.e., from 25 to 30, t reduces to 3.65. In case where the buying firm explores internally regarding the holding cost, it may be an advantage.

As the objective is to have a similar value of t, i.e., 4, the revised value of H, i.e., H_m can be obtained as follows:

$$t = \sqrt{\frac{2(C+I)}{H(D/T)}}$$

$$\text{Or } 4 = \sqrt{\frac{2 \times 400}{H_m \times 30}}$$

Or $H_m = ₹1.67$

If it is possible for the buying firm to reduce the H value to ₹1.67 from ₹2.00, the problem can be solved internally and the procurement cycle time is restored. By focusing on the storage as well as the cost of capital among other aspects, a reduction in H value may also be implemented.

Such a reduction in H can be generalized too. Let

$L = \%$ increase in $\left(\dfrac{D}{T}\right)$

$M = \%$ reduction in H

$$\sqrt{\frac{2(C+I)}{H(D/T)}} = \sqrt{\frac{2(C+I)}{H(1-M/100)(D/T)(1+L/100)}}$$

$$\text{Or } (1-M/100)(1+L/100) = 1$$

$$\text{Or } 1 - \frac{M}{100} = \frac{1}{(1+L/100)}$$

$$\text{Or } \frac{M}{100} = 1 - \frac{1}{(1+L/100)}$$

$$\text{Or } \frac{M}{100} = \frac{(L/100)}{(1+L/100)}$$

$$\text{Or } M = \frac{L}{(1+L/100)}$$

Table 2.4 shows the values of M for different levels of L.

In a reverse scenario, t increases with a reduction in H. In order to restore t, $\left(\dfrac{D}{T}\right)$ might be increased if it is possible for better conversion of potential buyers into the confirmed ones among other approaches.

Now,

$L = \%$ reduction in H

$M = \%$ increase in $\left(\dfrac{D}{T}\right)$

TABLE 2.4

Corresponding M (reduced H)

S. No.	L	M
1	3	2.91
2	6	5.67
3	9	8.26
4	12	10.71

TABLE 2.5

Corresponding $M\left\{\text{increased}\left(\dfrac{D}{T}\right)\right\}$

S. No.	L	M
1	3	3.09
2	6	6.38
3	9	9.89
4	12	13.64

$$\sqrt{\frac{2(C+I)}{H(D/T)}} = \sqrt{\frac{2(C+I)}{H(1-L/100)(D/T)(1+M/100)}}$$

Or $(1+M/100)(1-L/100)=1$

Or $1+\dfrac{M}{100} = \dfrac{1}{(1-L/100)}$

Or $\dfrac{M}{100} = \dfrac{(L/100)}{(1-L/100)}$

Or $M = \dfrac{L}{(1-L/100)}$

Table 2.5 shows the values of M for different levels of L.

Values of M are higher than that of L. Similarly, in Table 2.4, M values are higher, indicating relatively more efforts depending on the operational factors of the respective organization. This is verified as follows:

$$\frac{L}{(1-L/100)} > \frac{L}{(1+L/100)}$$

Or $(1+L/100) > (1-L/100)$

2.4.4   EXAMPLE WITH $\left(\dfrac{D}{T}\right)$ REDUCTION

Procurement cycle time increases with a reduction in the $\left(\dfrac{D}{T}\right)$ ratio. However, the supplying firm might find it difficult to adjust to this. In order to avoid unnecessary communication and coordination efforts, the buying firm can explore internally concerning the variation in suitable factors.

Consider the basic example data:

$$t = \sqrt{\frac{2(C+I)}{H(D/T)}}$$

$$\text{Or } t = \sqrt{\frac{2\times 400}{2\times 25}}$$

$$= 4 \text{ months}$$

When $\left(\dfrac{D}{T}\right)$ reduces from 25 to 20, the procurement cycle time increases to 4.47 months. In order to make it equivalent to 4 months, either the ordering cost or apportioned innovation cost or both might be reduced if it is convenient. For instance, while a trading firm sells an innovative product, the related innovation cost can depend on the following:

(i)   Identification of potential buyers
(ii) Appropriate efforts in convincing the potential buyer
(iii) Realization of a final billing in a suitable time

With a focus on such activities among others, it might be possible to reduce the related innovation cost.

Now,

$$4 = \sqrt{\frac{2(200 + I_m)}{2\times 20}}$$

where I_m is the reduced apportioned innovation cost.

Or $I_m = ₹120$

In order to generalize:

$L = \%$ reduction in $\left(\dfrac{D}{T}\right)$

$M = \%$ reduction in I

Also,

$$\sqrt{\frac{2(C+I)}{H(D/T)}} = \sqrt{\frac{2\{C + I(1 - M/100)\}}{H(D/T)(1 - L/100)}}$$

TABLE 2.6

Corresponding M (I) for $L\left(\dfrac{D}{T}\right)$

S. No.	L	M
1	5	10
2	10	20
3	15	30

$$\text{Or } (C+I) = \frac{C+I(1-M/100)}{(1-L/100)}$$

$$\text{Or } C(1-L/100) + I(1-L/100) = C + I(1-M/100)$$

$$\text{Or } I(1-M/100) = I(1-L/100) - (CL/100)$$

$$\text{Or } 1 - \frac{M}{100} = \frac{I(1-L/100) - (CL/100)}{I}$$

$$\text{Or } \frac{M}{100} = \frac{I - I(1-L/100) + (CL/100)}{I}$$

$$\text{Or } \frac{M}{100} = \frac{(IL/100) + (CL/100)}{I}$$

$$\text{Or } M = \frac{IL + CL}{I}$$

$$\text{Or } M = \frac{L(I+C)}{I}$$

$$\text{Or } M = L\left[1 + \frac{C}{I}\right] \tag{2.6}$$

Table 2.6 represents the corresponding values of M for various levels of L, while considering the example data.

Values of M are twice those of L because $C = ₹200$ and $I = ₹200$. Thus,

$$M = 2L$$

2.4.5 IMPLICATIONS OF $\left(\dfrac{C}{I}\right)$

When values of I are more than those of C, it is of interest to consider the reduced values of $\left(\dfrac{C}{I}\right)$. Table 2.7 gives the interaction of $\left(\dfrac{C}{I}\right)$ with the values of L and M.

TABLE 2.7

Interaction of $\left(\dfrac{C}{I}\right)$ with the Values of L and M

S. No.	1	2	3	4	5
$\left(\dfrac{C}{I}\right)$	0.9	0.8	0.7	0.6	0.5
M for $L = 5$	9.5	9.0	8.5	8.0	7.5
M for $L = 10$	19.0	18.0	17.0	16.0	15.0
M for $L = 15$	28.5	27.0	25.5	24.0	22.5

TABLE 2.8
M Values for Various Levels of N ($I > C$)

S. No.	N	M
1	10	19.09
2	20	18.33
3	30	17.69
4	40	17.14
5	50	16.67

In order to generalize, let

$$I = C\left(1 + \frac{N}{100}\right)$$

where the value of I is $N\%$ higher than that of C. From Eq. (2.6),

$$M = L\left[1 + \frac{C}{C(1 + N/100)}\right]$$

$$\text{Or } M = L\left[1 + \frac{1}{(1 + N/100)}\right] \tag{2.7}$$

For $L = 10$, the obtained values of M corresponding to various levels of N are given in Table 2.8.

In case where the organization is able to reduce the innovation efforts cost, let

$$I = C\left(1 - \frac{N}{100}\right)$$

TABLE 2.9
M Values for Various
Levels of N (I < C)

S. No.	N	M
1	10	21.11
2	20	22.50
3	30	24.29
4	40	26.67
5	50	30.00

where value of I is $N\%$ lower than that of C. From Eq. (2.6),

$$M = L\left[1 + \frac{C}{C(1 - N/100)}\right]$$

$$\text{Or } M = L\left[1 + \frac{1}{(1 - N/100)}\right] \tag{2.8}$$

For $L = 10$, the obtained values of M corresponding to various levels of N are provided in Table 2.9.

Compare with Table 2.8. The present values of M are higher than the previous scenario. It can be verified with reference to Eqs. (2.7) and (2.8) as follows:

$$1 + \frac{1}{(1 - N/100)} > 1 + \frac{1}{(1 + N/100)}$$

$$\text{Or } \frac{1}{(1 - N/100)} > \frac{1}{(1 + N/100)}$$

$$\text{Or } (1 + N/100) > (1 - N/100)$$

And this is true.

2.4.6 INTERACTION OF H AND I

In case where the value of H reduces because of certain favorable reasons, the value of t increases. For example, t is equivalent to 4 months in the basic example. Now with $H = ₹1.5$, the value of t is calculated as follows:

$$t = \sqrt{\frac{2(C + I)}{H(D/T)}}$$

$$\text{Or } t = \sqrt{\frac{2 \times 400}{1.5 \times 25}}$$

$$= 4\sqrt{\frac{4}{3}} \text{ months}$$

In order to restore the 4 months as t, either C or I, or both can be reduced if it is feasible. For example, if an effort is made to decrease I, then the reduced I, i.e., I_m is obtained as follows:

$$4 = \sqrt{\frac{2(200 + I_m)}{1.5 \times 25}}$$

Or $I_m = ₹100$

In order to generalize,
$L = \%$ reduction in H
$M = \%$ reduction in I
Also,

$$\sqrt{\frac{2(C + I)}{H(D/T)}} = \sqrt{\frac{2\left[C + I(1 - M/100)\right]}{H(1 - L/100)(D/T)}}$$

$$\text{Or } C + I = \frac{C + I(1 - M/100)}{(1 - L/100)}$$

$$\text{Or } C(1 - L/100) - C = I(1 - M/100) - I(1 - L/100)$$

$$\text{Or } -(CL/100) = -(IM/100) + (IL/100)$$

$$\text{Or } (IM/100) = (IL/100) + (CL/100)$$

$$\text{Or } (IM/100) = (I + C)(L/100)$$

$$\text{Or } IM = (I + C)L$$

$$\text{Or } M = L\left(1 + \frac{C}{I}\right)$$

While reducing I alone, this reduction is 50%, i.e., from ₹200 to ₹100. If the company feels that it is not possible, then both C and I can be reduced. For example, if it is not feasible to decrease I by more than 25%, then
$M = 25$
$N = \%$ reduction in C

Now,

$$\sqrt{\frac{2(C+I)}{H(D/T)}} = \sqrt{\frac{2\left[C(1-N/100)+I(1-M/100)\right]}{H(1-L/100)(D/T)}}$$

Or $(C+I)(1-L/100) = C(1-N/100)+I(1-M/100)$

Or $C(1-N/100) = (C+I)(1-L/100)-I(1-M/100)$

Or $1-\dfrac{N}{100} = \dfrac{(C+I)(1-L/100)-I(1-M/100)}{C}$

Or $\dfrac{N}{100} = \dfrac{C-(C+I)(1-L/100)+I(1-M/100)}{C}$

Or $N = 100\left[\dfrac{C-(C+I)(1-L/100)+I(1-M/100)}{C}\right]$ (2.9)

For $L = 25$, $M = 25$, $C = ₹200$, and $I = ₹200$, with the use of Eq. (2.9),
$N = 25$

Various combinations of M and N can also be generated in order to have certain flexibility in implementation, for a given value of L.

For example, $L = 25$, $C = ₹200$, and $I = ₹200$, with the use of Eq. (2.9),

$$N = 50 - M$$

Using the above expression, different combinations of M and N are shown in Table 2.10.

In another situation, i.e., the reverse scenario, I value may get reduced with certain effort. For restoring t, H value can be reduced if it is possible. In the basic example, if I value is reduced from ₹200 to ₹180 (i.e., 10% decrease), then the reduced H value, i.e., H_m, can be calculated as follows:

TABLE 2.10
M and N Combinations

S. No.	M	N
1	10	40
2	20	30
3	30	20
4	40	10

$$4 = \sqrt{\frac{2(200+180)}{25H_m}}$$

Or $H_m = ₹1.90$

For restoring t, the value of H needs to be decreased by 5%.

In order to generalize, let

L = % reduction in I

M = % reduction in H

Now,

$$\sqrt{\frac{2(C+I)}{H(D/T)}} = \sqrt{\frac{2[C+I(1-L/100)]}{H(1-M/100)(D/T)}}$$

Or $(C+I)(1-M/100) = C+I(1-L/100)$

Or $1 - \dfrac{M}{100} = \dfrac{C+I(1-L/100)}{(C+I)}$

Or $M = 100 \left[\dfrac{C+I-C-I(1-L/100)}{(C+I)} \right]$

Or $M = 100 \left[\dfrac{(IL/100)}{(C+I)} \right]$

Or $M = \dfrac{IL}{(C+I)}$

Or $M = \dfrac{L}{1+(C/I)}$ (2.10)

Table 2.11 gives the M values corresponding to that of L for the given data.

TABLE 2.11
M (for a Reduced H)
Corresponding to L
(for a Reduced I)

S. No.	L	M
1	5	2.5
2	10	5.0
3	15	7.5
4	20	10.0

Values of M are obtained as half of L because of $C = ₹200$ and $I = ₹200$. Thus,

$$M = \frac{L}{2}$$

Now, the role of $\left(\dfrac{C}{I}\right)$ ratio can be understood. When the value of I is more than that of C, it is of interest to consider the reduced values of $\left(\dfrac{C}{I}\right)$. Table 2.12 gives the interaction of $\left(\dfrac{C}{I}\right)$ with the values of L and M.

In order to generalize, let

$$I = C\left(1 + \frac{N}{100}\right)$$

where the value of I is $N\%$ higher than that of C. From Eq. (2.10),

$$M = \frac{L}{\left[1 + \dfrac{C}{C(1 + N/100)}\right]}$$

$$\text{Or } M = \frac{L}{\left[1 + \dfrac{1}{(1 + N/100)}\right]} \tag{2.11}$$

For $L = 10$, the obtained values of M corresponding to various levels of N are given in Table 2.13.

In case where the organization is able to reduce the innovation efforts cost, let

$$I = C\left(1 - \frac{N}{100}\right)$$

where value of I is $N\%$ lower than that of C. From Eq. (2.10),

---

TABLE 2.12

Interaction of $\left(\dfrac{C}{I}\right)$ **with the Values of L and M**

S. No.	1	2	3	4	5
$\left(\dfrac{C}{I}\right)$	0.9	0.8	0.7	0.6	0.5
M for $L = 5$	2.63	2.78	2.94	3.12	3.33
M for $L = 10$	5.26	5.56	5.88	6.25	6.67
M for $L = 15$	7.89	8.33	8.82	9.38	10.00

TABLE 2.13

M Values for Various Levels of N (I > C)

S. No.	N	M
1	10	5.24
2	20	5.45
3	30	5.65
4	40	5.83
5	50	6.00

TABLE 2.14

M Values for Various Levels of N (I < C)

S. No.	N	M
1	10	4.74
2	20	4.44
3	30	4.12
4	40	3.75
5	50	3.33

$$M = \frac{L}{\left[1 + \dfrac{C}{C(1 - N/100)}\right]}$$

$$\text{Or } M = \frac{L}{\left[1 + \dfrac{1}{(1 - N/100)}\right]} \tag{2.12}$$

For $L = 10$, the obtained values of M corresponding to various levels of N are provided in Table 2.14.

Compare with Table 2.13. The present values of M are lower than those in the previous scenario. It can be verified with reference to Eqs. (2.11) and (2.12) as follows:

$$\frac{1}{\left[1 + \dfrac{1}{(1 - N/100)}\right]} < \frac{1}{\left[1 + \dfrac{1}{(1 + N/100)}\right]}$$

$$\text{Or } 1 + \frac{1}{(1 + N/100)} < 1 + \frac{1}{(1 - N/100)}$$

$$\text{Or} \quad \frac{1}{(1+N/100)} < \frac{1}{(1-N/100)}$$

$$\text{Or} \quad (1-N/100) < (1+N/100)$$

And this is true.

In these procurement examples, certain conclusions are drawn numerically as well as analytically. The influence of variation in a factor is studied and it is shown how the procurement cycle time varies. It is of practical and business interest to restore this cycle time. A remedial measure lies in the multiple organizational factors and their change if it is feasible and convenient from the operational and managerial point of view, i.e., inclusive of qualitative as well as quantitative aspects related to the problem.

2.5   IMPERFECT LOT

A procurement lot might be of two kinds, as shown in Figure 2.8.

In a perfect lot, the whole quantity ordered is acceptable from the quality point of view. However, in an imperfect lot, there are certain unacceptable items in a purchased quantity. For instance, consider that 900 components are required by a manufacturing firm in 36 months of a planned duration from a supplier company. And it is estimated that the proportion of acceptable components is 0.9, depending on the purchase experience. In such a case, the total requirement would be

$$\frac{900}{0.9} = 1000$$

Thus, the total number of components ordered for procurement can be given as follows:

$$\frac{D}{y}$$

where y = proportion of acceptable components in a lot.

Now, two situations can arise. In the first case, the procured quantity in each cycle is adjusted by this proportion, as shown in Figure 2.9; however, the unacceptable components are not added to the inventory.

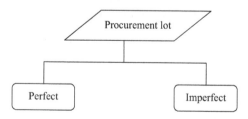

FIGURE 2.8   Perfect and imperfect lot.

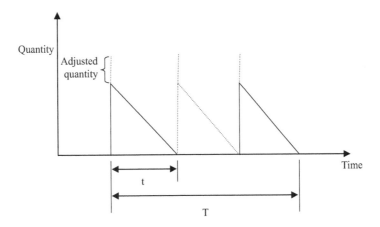

FIGURE 2.9 Imperfect lot with no addition to inventory.

This can happen when the unacceptable components are identified as soon as these are received, and are not added to the inventory. Thus, the inventory carrying costs are not affected. However, the purchased quantity in each cycle needs to be adjusted with a certain proportion and that may impact the purchase cost.

In the second situation, it may not be possible to identify the unacceptable components instantaneously on receipt, and these are added to the inventory level. This affects inventory costs as well as the procurement cycle time. Now,

Procurement quantity per cycle $= \dfrac{Dt}{yT}$

Average quantity during the cycle $= \dfrac{Dt}{2yT}$

Inventory cost during the cycle $= \dfrac{Dt}{2yT} \cdot H \cdot t$

Total inventory cost for the planned duration $(T) = \dfrac{Dt}{2yT} \cdot Ht \cdot \dfrac{T}{t}$

$$= \frac{DtH}{2y} \tag{2.13}$$

After adding the ordering and innovation cost, the estimated total cost (E) can be expressed as follows:

$$E = \frac{DtH}{2y} + \frac{T}{t} \cdot (C+I) \tag{2.14}$$

Optimal value of t can be obtained by differentiating Eq. (2.14) with respect to t and equating to zero:

$$\frac{dE}{dt} = \frac{DH}{2y} - \frac{T(C+I)}{t^2} = 0$$

$$\text{Or} \quad \frac{T(C+I)}{t^2} = \frac{DH}{2y}$$

$$\text{Or} \quad t^2 = \frac{2Ty(C+I)}{DH}$$

$$\text{Or} \quad t = \sqrt{\frac{2Ty(C+I)}{DH}} \qquad\qquad (2.15)$$

In order to illustrate, consider the following example:

$T = 36$ months
$D = 900$ number of components
$H = ₹2$
$C = ₹200$
$I = ₹200$
$y = 0.81$

The procurement cycle time can be obtained with the use of Eq. (2.15) as
$t = 3.6$ months
Procurement cycle time increases with the increase in y, as shown in Table 2.15.
However, a proportional increase in y may be a suitable factor for a generalized approach in order to know the variation in t.
Let
$L = \%$ increase in y
$t_m = $ value of t with $\%$ variation in y

$$\% \text{ increase in } t = 100\left(\frac{t_m - t}{t}\right)$$

TABLE 2.15
Variation of t with y

S. No.	y	T
1	0.81	3.60
2	0.84	3.67
3	0.87	3.73
4	0.90	3.79
5	0.93	3.86
6	0.96	3.92

$$= 100\left(\frac{t_m}{t} - 1\right)$$

where $t_m = \sqrt{\dfrac{2Ty(1+L/100)(C+I)}{DH}}$

and $t = \sqrt{\dfrac{2Ty(C+I)}{DH}}$

Therefore, the % increase in $t = 100\left[\sqrt{(1+L/100)} - 1\right]$

Table 2.16 shows the percentage increase in t.

Now,

% reduction in $t = 100\left(\dfrac{t - t_m}{t}\right)$

$$= 100\left(1 - \frac{t_m}{t}\right)$$

where $t_m = \sqrt{\dfrac{2Ty(1-L/100)(C+I)}{DH}}$

$L = $ % reduction in y

$$t = \sqrt{\dfrac{2Ty(C+I)}{DH}}$$

Therefore, the % reduction in $t = 100\left[1 - \sqrt{(1-L/100)}\right]$

Table 2.17 shows the percentage reduction in t.

On comparison with the previous situation, the percentage variation in t is relatively higher. This can be proved as follows:

$$1 - \sqrt{(1-L/100)} > \sqrt{(1+L/100)} - 1$$

TABLE 2.16

Percentage Increase in t with Respect to y

S. No.	L	% Increase in t
1	3	1.49
2	6	2.96
3	9	4.40
4	12	5.83
5	15	7.24

TABLE 2.17

Percentage Reduction in t

S. No.	L	% Reduction in t
1	3	1.51
2	6	3.05
3	9	4.61
4	12	6.19
5	15	7.80

$$\text{Or } \sqrt{(1+L/100)} + \sqrt{(1-L/100)} < 2$$

$$\text{Or } 1+(L/100)+1-(L/100)+2\sqrt{1-(L/100)^2} < 4$$

$$\text{Or } 1+\sqrt{1-(L/100)^2} < 2$$

$$\text{Or } \sqrt{1-(L/100)^2} < 1$$

L is less than 100 for all practical purposes, and hence it is verified.

2.5.1 DEALING WITH THE *t* INCREASE

When y value increases, the purchase cycle time t also increases in case where unacceptable components are added to the inventory. In view of the significance of restoring the t value discussed before, it can be explored. In order to restore the cycle time, an innovation cost may be reduced if it is possible. Now,

L = % increase in y
M = % reduction in I

With the use of Eq. (2.15),

$$\sqrt{\frac{2Ty(C+I)}{DH}} = \sqrt{\frac{2Ty(1+L/100)\{C+I(1-M/100)\}}{DH}}$$

$$\text{Or } (C+I) = (1+L/100)\{C+I(1-M/100)\}$$

$$\text{Or } \frac{(C+I)}{(1+L/100)} = C+I(1-M/100)$$

$$\text{Or } I(1-M/100) = \frac{(C+I)}{(1+L/100)} - C$$

$$\text{Or } I(1 - M/100) = \frac{C + I - C(1 + L/100)}{(1 + L/100)}$$

$$\text{Or } I(1 - M/100) = \frac{I - (CL/100)}{(1 + L/100)}$$

$$\text{Or } 1 - \frac{M}{100} = \frac{I - (CL/100)}{I(1 + L/100)}$$

$$\text{Or } \frac{M}{100} = \frac{I(1 + L/100) - I + (CL/100)}{I(1 + L/100)}$$

$$\text{Or } \frac{M}{100} = \frac{(IL/100) + (CL/100)}{I(1 + L/100)}$$

$$\text{Or } M = \frac{(IL + CL)}{I(1 + L/100)}$$

$$\text{Or } M = \frac{L(I + C)}{I(1 + L/100)}$$

$$\text{Or } M = \frac{L}{(1 + L/100)}\left[1 + \frac{C}{I}\right] \tag{2.16}$$

Considering the relevant example data,

$C = ₹200$
$I = ₹200$

Table 2.18 represents the corresponding values of M for various levels of L when

$$\frac{C}{I} = 1$$

When values of I are more than that of C, it is of interest to consider the reduced values of $\left(\dfrac{C}{I}\right)$. Table 2.19 gives the interaction of $\left(\dfrac{C}{I}\right)$ with the values of M for $L = 10$.

TABLE 2.18
Corresponding
M (I) for L (y)

S. No.	L	M
1	5	9.52
2	10	18.18
3	15	26.09

TABLE 2.19

Interaction of $\left(\dfrac{C}{I}\right)$ with the Values of M

S. No.	1	2	3	4	5
$\left(\dfrac{C}{I}\right)$	0.9	0.8	0.7	0.6	0.5
M for L = 10	17.27	16.36	15.45	14.55	13.64

TABLE 2.20
M Values for Various Levels of N (I > C)

S. No.	N	M
1	10	17.36
2	20	16.67
3	30	16.08
4	40	15.58
5	50	15.15

In order to generalize, let

$$I = C\left(1 + \frac{N}{100}\right)$$

where the value of I is $N\%$ higher than that of C. From Eq. (2.16),

$$M = \frac{L}{(1 + L/100)}\left[1 + \frac{1}{(1 + N/100)}\right]$$

For $L = 10$, the obtained values of M corresponding to various levels of N are given in Table 2.20.

In case where the organization is able to reduce the innovation efforts cost, let

$$I = C\left(1 - \frac{N}{100}\right)$$

where the value of I is $N\%$ lower than that of C. From Eq. (2.16),

$$M = \frac{L}{(1 + L/100)}\left[1 + \frac{1}{(1 - N/100)}\right]$$

TABLE 2.21

M Values for Various Levels of N (I < C)

S. No.	N	M
1	10	19.19
2	20	20.45
3	30	22.08
4	40	24.24
5	50	27.27

For $L = 10$, the obtained values of M corresponding to various levels of N are provided in Table 2.21.

Comparing with the previous scenario, the present values of M are higher. This is because

$$(1 - N / 100) < (1 + N / 100)$$

2.5.2 DEALING WITH THE *t* REDUCTION

When the y value reduces, the purchase cycle time t also reduces in the present context. In order to restore the cycle time, a holding cost may be reduced if it is possible. In order to illustrate,

$T = 36$ months
$D = 900$ number of components
$H = ₹2$
$C = ₹200$
$I = ₹200$
$y = 0.96$

From Eq. (2.15),
$t = 3.92$ months
When y value reduces to 0.90, then
$t = 3.79$ months
In order to restore the t value, the revised value of H, i.e., H_m can be obtained as follows:

$$t = \sqrt{\frac{2Ty(C+I)}{DH}}$$

$$\text{Or } 3.92 = \sqrt{\frac{2 \times 36 \times 0.90 \times 400}{900 H_m}}$$

Or $H_m = ₹1.87$

A rigorous analysis is done for the perfect as well as imperfect lot in the context of procurement situation.

2.6   PERMISSIBLE BACKORDERS

When shortages are permitted and these are completely backordered, such a situation is represented by Figure 2.10, where

J = maximum permissible shortages which are totally backordered

Now,

Procurement quantity per cycle $= \dfrac{Dt}{T}$

Average quantity during the cycle for positive inventory $= \dfrac{1}{2}\left(\dfrac{Dt}{T} - J\right)$

As the time corresponding to the positive inventory in each cycle is $t - \dfrac{JT}{D}$,

Inventory cost during the cycle $= \dfrac{1}{2}\left(\dfrac{Dt}{T} - J\right) \cdot H\left(t - \dfrac{JT}{D}\right)$

Total inventory cost $= \dfrac{1}{2}\left(\dfrac{Dt}{T} - J\right) \cdot H\left(t - \dfrac{JT}{D}\right) \cdot \dfrac{T}{t}$

$$= \dfrac{1}{2}\left(D - \dfrac{JT}{t}\right) \cdot H\left(t - \dfrac{JT}{D}\right) \qquad (2.17)$$

Time in each cycle when stock out happens $= \dfrac{JT}{D}$

Average stock out quantity $= \dfrac{J}{2}$

Backordering or stock out cost per cycle $= \dfrac{J}{2}\left(\dfrac{JT}{D}\right) \cdot K$

where K = stock out cost per unit for one month

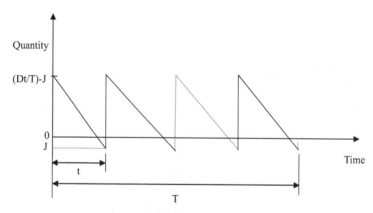

FIGURE 2.10   Permissible shortages.

$$\text{Total stock out cost} = \frac{J}{2}\left(\frac{JT}{D}\right) \cdot K \cdot \frac{T}{t}$$

$$= \left(\frac{JT}{D}\right)\left(\frac{JKT}{2t}\right) \tag{2.18}$$

$$\text{Total ordering and innovation cost} = \frac{T}{t}(C+I) \tag{2.19}$$

Adding Eqs. (2.17), (2.18), and (2.19), the total relevant cost can be obtained as follows:

$$E = \frac{H}{2}\left(D - \frac{JT}{t}\right)\left(t - \frac{JT}{D}\right) + \left(\frac{JT}{D}\right)\left(\frac{JKT}{2t}\right) + \frac{T}{t}(C+I)$$

$$\text{Or } E = \frac{H}{2}\left[Dt - JT - JT + \frac{J^2T^2}{Dt}\right] + \frac{J^2T^2K}{2Dt} + \frac{T}{t}(C+I)$$

$$\text{Or } E = \frac{HDt}{2} - JTH + \frac{J^2T^2H}{2Dt} + \frac{J^2T^2K}{2Dt} + \frac{T}{t}(C+I)$$

$$\text{Or } E = \frac{HDt}{2} - JTH + \frac{J^2T^2}{2Dt}(H+K) + \frac{T}{t}(C+I) \tag{2.20}$$

$\dfrac{\partial E}{\partial J} = 0$ shows

$$-TH + \frac{JT^2}{Dt}(H+K) = 0$$

$$\text{Or } J = \frac{HDt}{T(H+K)} \tag{2.21}$$

Using Eq. (2.20), $\dfrac{\partial E}{\partial t} = 0$ shows

$$\frac{HD}{2} - \frac{J^2T^2(H+K)}{2Dt^2} - \frac{T(C+I)}{t^2} = 0$$

$$\text{Or } \frac{HD}{2} = \frac{1}{t^2}\left[\frac{J^2T^2(H+K)}{2D} + T(C+I)\right]$$

Substituting the value of J from Eq. (2.21),

$$\frac{HD}{2} = \frac{1}{t^2}\left[\frac{H^2Dt^2}{2(H+K)} + T(C+I)\right]$$

$$\text{Or } \frac{HD}{2} = \frac{H^2D}{2(H+K)} + \frac{T(C+I)}{t^2}$$

$$\text{Or } \frac{HD}{2}\left[1 - \frac{H}{(H+K)}\right] = \frac{T(C+I)}{t^2}$$

$$\text{Or } \frac{HD}{2}\left[\frac{K}{(H+K)}\right] = \frac{T(C+I)}{t^2}$$

$$\text{Or } t^2 = \frac{2T(H+K)(C+I)}{HDK}$$

$$\text{Or } t = \sqrt{\frac{2T(H+K)(C+I)}{HDK}} \qquad\qquad (2.22)$$

In order to illustrate, consider the following example:

$T = 36$ months
$D = 864$ number of components
$H = ₹2$
$C = ₹185$
$I = ₹185$
$K = ₹50$

The procurement cycle time can be obtained with the use of Eq. (2.22) as

$t = 4$ months

Procurement cycle time increases with a reduction in K, as shown in Table 2.22.
    However, a proportional reduction in K may be a suitable factor for a generalized approach in order to know the variation in t.
    Let
$L = \%$ reduction in K
$t_m =$ value of t with $\%$ variation in K

$$\% \text{ increase in } t = 100\left(\frac{t_m - t}{t}\right)$$

$$= 100\left(\frac{t_m}{t} - 1\right)$$

TABLE 2.22

Variation of *t* with *K*

S. No.	K	T
1	45	4.01
2	40	4.02
3	35	4.04
4	30	4.06

TABLE 2.23

Percentage Increase in *t* with Respect to *K*

S. No.	L	% Increase in t
1	8	0.17
2	11	0.24
3	14	0.31
4	17	0.39
5	20	0.48

where $t_m = \sqrt{\dfrac{2T(C+I)\{H+K(1-L/100)\}}{HDK(1-L/100)}}$

and $t = \sqrt{\dfrac{2T(H+K)(C+I)}{HDK}}$

Therefore, the % increase in $t = 100\left[\sqrt{\dfrac{H+K(1-L/100)}{(1-L/100)(H+K)}}-1\right]$

Table 2.23 shows the percentage increase in t.

In order to restore the t value, the innovation cost may be reduced if it is feasible. Now,

$L = \%$ reduction in K

$M = \%$ reduction in I

$$\sqrt{\dfrac{2T(H+K)(C+I)}{HDK}}=\sqrt{\dfrac{2T\{H+K(1-L/100)\}\{C+I(1-M/100)\}}{HDK(1-L/100)}}$$

Or $\dfrac{(H+K)(1-L/100)(C+I)}{\{H+K(1-L/100)\}}=C+I(1-M/100)$

$$\text{Or } 1 - \frac{M}{100} = \frac{(H+K)(1-L/100)(C+I) - C\{H + K(1-L/100)\}}{I\{H + K(1-L/100)\}}$$

$$\text{Or } \frac{M}{100} = \frac{I\{H + K(1-L/100)\} - (H+K)(1-L/100)(C+I) + C\{H + K(1-L/100)\}}{I\{H + K(1-L/100)\}}$$

$$\text{Or } \frac{M}{100} = \frac{(C+I)\{H + K(1-L/100)\} - (H+K)(1-L/100)(C+I)}{I\{H + K(1-L/100)\}}$$

$$\text{Or } \frac{M}{100} = \frac{(C+I)\{H + K(1-L/100) - H(1-L/100) - K(1-L/100)\}}{I\{H + K(1-L/100)\}}$$

$$\text{Or } \frac{M}{100} = \frac{(C+I)\{H - H + (HL/100)\}}{I\{H + K(1-L/100)\}}$$

$$\text{Or } M = 100\left[\frac{(C+I)(HL/100)}{I\{H + K(1-L/100)\}}\right]$$

$$\text{Or } M = \frac{(C+I)HL}{I\{H + K(1-L/100)\}}$$

Table 2.24 shows the corresponding M for various levels of L with the given data. Now,

$$\% \text{ reduction in } t = 100\left(\frac{t - t_m}{t}\right)$$

$$= 100\left(1 - \frac{t_m}{t}\right)$$

TABLE 2.24

Corresponding M (I) for L (Reduced K)

S. No.	L	M
1	8	0.67
2	11	0.95
3	14	1.24
4	17	1.56
5	20	1.90

where $t_m = \sqrt{\dfrac{2T(C+I)\{H+K(1+L/100)\}}{HDK(1+L/100)}}$

$L = \%$ increase in K

and $t = \sqrt{\dfrac{2T(H+K)(C+I)}{HDK}}$

Therefore, the % reduction in $t = 100\left[1 - \sqrt{\dfrac{H+K(1+L/100)}{(1+L/100)(H+K)}}\right]$

Table 2.25 shows the percentage reduction in t.

In order to restore the t value, the holding cost may be explored for reduction:

$L = \%$ increase in K

$M = \%$ reduction in H

$$\sqrt{\dfrac{2T(H+K)(C+I)}{HDK}} = \sqrt{\dfrac{2T\{H(1-M/100)+K(1+L/100)\}(C+I)}{H(1-M/100)DK(1+L/100)}}$$

Or $(H+K)(1-M/100)(1+L/100) = H(1-M/100)+K(1+L/100)$

Or $(1-M/100)\{(H+K)(1+L/100)-H\} = K(1+L/100)$

Or $(1-M/100) = \dfrac{K(1+L/100)}{H(1+L/100)+K(1+L/100)-H}$

Or $(1-M/100) = \dfrac{K(1+L/100)}{(HL/100)+K(1+L/100)}$

Or $\dfrac{M}{100} = \dfrac{(HL/100)+K(1+L/100)-K(1+L/100)}{(HL/100)+K(1+L/100)}$

TABLE 2.25

Percentage Reduction in t with Respect to K

S. No.	L	% Reduction in t
1	8	0.14
2	11	0.19
3	14	0.24
4	17	0.28
5	20	0.32

TABLE 2.26

Corresponding M (H) for L (increased K)

S. No.	L	M
1	8	0.30
2	11	0.39
3	14	0.49
4	17	0.58
5	20	0.66

Or $$\frac{M}{100} = \frac{(HL/100)}{(HL/100) + K(1 + L/100)}$$

Or $$M = \frac{HL}{K + (H + K)(L/100)}$$

Table 2.26 shows the corresponding M for various levels of L with the given data.

Various aspects pertaining to the procurement inventory have been discussed and analyzed rigorously, including the permissible shortages that are totally backordered in the system.

3 Production Inventory

Manufacture of a product involves various processes on a component or input item. Production inventory refers to the manufactured component or product. Innovation efforts are also included along with other aspects of the manufactured inventory. In order to link the innovation concept and an associated cost with the production inventory, it is necessary to understand the manufacture in the present discussion.

3.1 MANUFACTURE

Resources are consumed in the manufacture of a product, as shown in Figure 3.1, where the types of resources are numerous.

These resources might include the production or manufacturing facilities and human resources among others. The facilities are installed depending on the layout, available space, and the sequence of processes that are to be carried out on the corresponding input items or components. In the basic manufacture, either input item/raw material is converted to a component, or a component/subassembly is converted to a useful product/finished assembly. The number of production facilities may vary as per the manufacturing need. However, either one or a few facilities may also be considered depending on the scope of the problem to be analyzed.

3.1.1 SINGLE PRODUCTION FACILITY

In a manufacturing environment, a component can be the output of a production shop or can be used as an intermediate resource for further processes. In order to produce a component from the raw material or an input item, single production facility might also be utilized as shown in Figure 3.2 in a certain specific case.

Production capacity of this single facility is an important aspect of the manufacturing firm. Production capacity is associated with certain time zones:

(a) Year
(b) Month
(c) Week
(d) Day
(e) Hour

Therefore, it is possible to know the produced number of components by such a facility in certain total time duration.

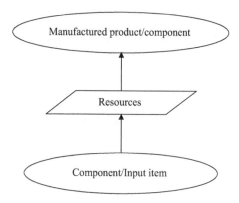

FIGURE 3.1   Basic manufacture.

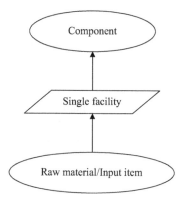

FIGURE 3.2   Single production facility.

3.1.2   GROUP OF PRODUCTION FACILITIES

In order to manufacture a useful product or finished assembly, a group of production facilities can be used, as shown in Figure 3.3. Installation of the facilities is done on the basis of many factors:

(i)   Type and nature of product
(ii)   Production volume
(iii)   Type of layout
(iv)   Available area for effective use
(v)   Type of facility/machine
(vi)   Material handling

However, multiple production facilities are shown in sequence in Figure 3.3 for convenience. Several processes are carried out one after another on the input item and transformed component. An overall production capacity associated with a certain

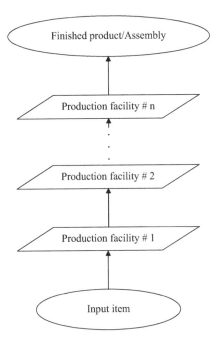

FIGURE 3.3   Multiple production facilities.

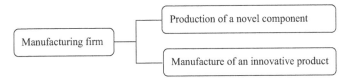

FIGURE 3.4   Possibilities before a manufacturing firm.

time zone finally at the last facility stage in the sequence might be of interest as per the scope of the problem and analysis. However, when a particular work-in-process is the focus area, any one or adjacent production facility can be identified from this group of facilities. Production inventory can be accumulated or in-process at any stage.

3.2   INNOVATION

Some relevant aspects pertaining to the innovation efforts are discussed next.

3.2.1   POSSIBILITIES FOR A PRODUCTION FIRM

A manufacturing firm can decide to produce an innovative item completely. However, it may also confine to the production of a novel component which would be used in an existing finished assembly. Such possibilities are shown in Figure 3.4.

TABLE 3.1

Manufacture of Innovative Component/Product

Component	Innovative Product
Limited design effort	Extensive design effort
Choice of material may be relatively convenient	Choice of material may be relatively cumbersome
Affects part of the process	May affect the production process significantly
Limited training of employees	Extensive training of employees
Coordination efforts on lower side	Coordination efforts on higher side

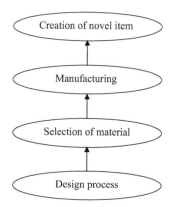

FIGURE 3.5   Sequential process.

Production of a novel component may be appropriate when this transforms the existing assembly into a promising innovative final product. Alternatively, the firm can engage in the creation of an innovative product completely. Differentiation in these approaches is summarized in Table 3.1.

3.2.2   SEQUENTIAL PROCESS

Innovative component/product has been mentioned before. In order to understand this differentiation, a sequential process of manufacture in the context of a novel item/component is shown in Figure 3.5.

As the final product/assembly is composed of many components, design effort may be visualized accordingly. Thus, a limited design effort corresponds to the innovative component, whereas an extensive effort is needed for the design of an innovative final product. In case where a substitution of material is required, a choice is to be made for one or very few materials for the component. On the other hand, the types of materials needed for the whole product might be many. Thus, the selection of material in the context of novelty might be relatively difficult and time consuming. Now the manufacturing activity on the chosen material starts involving various production processes on the facilities. The effect on production processes is significant in case of the whole product, whereas such an effect may also be on the part of the complete process in the case of the component.

Employees' training may be necessary for the following:

(i) Novel design
(ii) Substitution of material
(iii) Process design

Such training, if necessary, can be limited/extensive in the context of component/ product on the basis of inherent complexity. In all the above-mentioned activities, considerable coordination efforts are needed. Such efforts are relatively on the higher side with reference to the creation of whole product.

The related innovation efforts should be listed fully for the purpose of the total innovation cost determination. This total innovation cost can be apportioned for a cyclic approach to production. Such a conversion of efforts into the innovation cost helps in the total cost planning.

3.3 RELATED TOTAL COST PLANNING

Various cost components have been formulated in order to approach the related total cost for certain planned duration.

3.3.1 COST COMPONENTS

Consider one production cycle, as shown in Figure 3.6. Similar cycle gets repeated for a planned duration on the basis of a certain strategic period.

Consider an example where 900 units of demand an estimated over a period of 36 months with almost uniform consumption throughout. In a total period of 36 months, 900 units of an innovative product are to be manufactured.

$$\text{The demand per month} = \frac{900}{36}$$

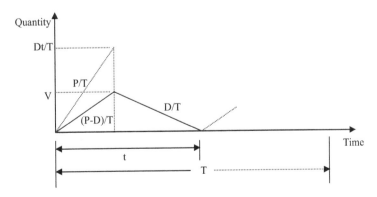

FIGURE 3.6 Production cycle and continuation for a planned horizon.

In case where a production cycle is of 4 months,

the production quantity per cycle $= \dfrac{900}{36} \times 4$

In order to generalize, refer to Figure 3.6 where

t = production cycle time in months
T = total planned duration in months

Now,

Production quantity per cycle $= \dfrac{Dt}{T}$

where D = total demand for the planned duration (T)

Average inventory during the cycle $= \dfrac{V}{2}$

where V = maximum production inventory during the cycle

Production inventory buildup happens during the production time only in a cycle. Thus,

$$\frac{V}{(P-D)/T} = \frac{(Dt/T)}{(P/T)}$$

where P = total production capacity for the total duration T

$$\text{Or } V = \frac{(1-D/P)Dt}{T} \tag{3.1}$$

On the basis of average inventory, the holding cost is estimated. Therefore,

Inventory holding cost for the production cycle $= \dfrac{V}{2} \cdot H \cdot t$

where H = inventory holding cost per product in ₹/month

With the substitution of Eq. (3.1),

The production inventory holding cost per cycle $= \dfrac{(1-D/P)Dt}{2T} \cdot H \cdot t$

As there are $\left(\dfrac{T}{t}\right)$ production cycles in the total planned duration,

$$\text{Total production inventory cost} = \frac{(1-D/P)Dt}{2T} \cdot Ht \cdot \frac{T}{t}$$

$$= \frac{(1-D/P)DHt}{2} \tag{3.2}$$

$$\text{Total innovation cost} = I \cdot \frac{T}{t} \tag{3.3}$$

where I = apportioned innovation cost per production cycle

$$\text{Total facility setup cost} = C \cdot \frac{T}{t} \qquad (3.4)$$

where C = fixed setup cost per facility setup

3.3.2 Production Cycle Time

Total cost (E) for the whole planned duration (T) can be obtained by adding Eqs. (3.2), (3.3), and (3.4):

$$E = \frac{(1 - D/P)DHt}{2} + (C + I)\frac{T}{t} \qquad (3.5)$$

In order to get the optimal production cycle time, differentiate with respect to t and equate to zero:

$$\frac{(1 - D/P)DH}{2} - \frac{(C + I)T}{t^2} = 0$$

$$\text{Or } t = \sqrt{\frac{2T(C + I)}{(1 - D/P)DH}} \qquad (3.6)$$

In order to illustrate, consider a basic example:

A total planned duration T = 36 months
Total demand D = 900
Total production capacity P = 1200
Facility setup cost C = ₹500
Apportioned innovation cost per cycle I = ₹500
Inventory holding cost per unit product H = ₹20 per month

Now,
From Eq. (3.6),

$$t = \sqrt{\frac{2 \times 36 \times (500 + 500)}{(1 - 900/1200) \times 900 \times 20}}$$

Or t = 4 months

3.4 PRODUCTION EXAMPLES

In the mentioned illustrative example, production cycle time t is obtained as 4 months. The innovation cost can vary depending on the specific environment related to the following:

(a) Investment level
(b) Sources of capital

(c) Organizational characteristics
(d) Innovative capability
(e) Human resources

Therefore, the variation in I and its impact needs to be analyzed so that the information can be generated for effect on production organization, including the cycle time. This information can be useful for feasible appropriate changes in operational factors.

3.4.1   VARIATION IN THE INNOVATION COST

An increase in I is first considered. Table 3.2 gives the production cycle time (t in months) corresponding to different I values with the use of Eq. (3.6) and the remaining data as follows:

$T = 36$ months
$C = ₹500$
$D = 900$
$P = 1200$
$H = ₹20$

Production cycle time increases with the higher values of I.
    In order to generalize,

$$\text{Varied } I_m = I\left(1 + \frac{L}{100}\right)$$

where $L = \%$ increase in I

    $\%$ increase in production cycle time, $M = 100\left(\frac{t_m - t}{t}\right)$

    where t_m = varied production cycle time

TABLE 3.2
Increased I Values and Corresponding t

S. No.	I	t
1	500	4.00
2	600	4.20
3	700	4.38
4	800	4.56
5	900	4.73

TABLE 3.3

Corresponding M for L (Increased I)

S. No.	L	M
1	20	4.88
2	40	9.54
3	60	14.02
4	80	18.32
5	100	22.47

Now,

$$M = 100\left(\frac{t_m}{t} - 1\right)$$

$$\text{Or } M = 100\left[\sqrt{\frac{2T(C + I_m)}{(1 - D/P)DH}} \cdot \sqrt{\frac{(1 - D/P)DH}{2T(C + I)}} - 1\right]$$

$$\text{Or } M = 100\left[\sqrt{\frac{(C + I_m)}{(C + I)}} - 1\right]$$

$$\text{Or } M = 100\left[\sqrt{\frac{C + I(1 + L/100)}{(C + I)}} - 1\right]$$

Corresponding values of M are shown in Table 3.3 for the given data, including $I = ₹500$, for different values of L.

Similarly, the % reduction in the production cycle time can be obtained with the % reduction in the I values.

Reduced value, $I_m = I\left(1 - \frac{L}{100}\right)$

where $L = \%$ reduction in I, and

% reduction in production cycle time, $M = 100\left(\frac{t - t_m}{t}\right)$

where $t_m =$ reduced production cycle time

Now,

$$M = 100\left(1 - \frac{t_m}{t}\right)$$

$$\text{Or } M = 100\left[1 - \sqrt{\frac{2T(C + I_m)}{(1 - D/P)DH}} \cdot \sqrt{\frac{(1 - D/P)DH}{2T(C + I)}}\right]$$

$$\text{Or } M = 100\left[1 - \sqrt{\frac{(C+I_m)}{(C+I)}}\right]$$

$$\text{Or } M = 100\left[1 - \sqrt{\frac{C+I(1-L/100)}{(C+I)}}\right]$$

Corresponding values of M are shown in Table 3.4 for the given data for different values of L.

It appears that the present values of M are higher than that corresponding to the previous scenario. This can be verified as follows:

$$100\left[1 - \sqrt{\frac{C+I(1-L/100)}{(C+I)}}\right] > 100\left[\sqrt{\frac{C+I(1+L/100)}{(C+I)}} - 1\right]$$

$$\text{Or } \sqrt{\frac{C+I(1+L/100)}{(C+I)}} + \sqrt{\frac{C+I(1-L/100)}{(C+I)}} < 2$$

$$\text{Or } \sqrt{C+I(1+L/100)} + \sqrt{C+I(1-L/100)} < 2\sqrt{(C+I)}$$

$$\text{Or } C+I(1+L/100) + C+I(1-L/100) + 2\sqrt{\{C+I(1+L/100)\}\{C+I(1-L/100)\}}$$
$$< 4(C+I)$$

$$\text{Or } 2C+2I + 2\sqrt{\{C+I(1+L/100)\}\{C+I(1-L/100)\}} < 4(C+I)$$

$$\text{Or } (C+I) + \sqrt{\{C+I(1+L/100)\}\{C+I(1-L/100)\}} < 2(C+I)$$

$$\text{Or } \sqrt{\{C+I(1+L/100)\}\{C+I(1-L/100)\}} < (C+I)$$

TABLE 3.4

Corresponding M for L (reduced I)

S. No.	L	M
1	10	2.53
2	20	5.13
3	30	7.80
4	40	10.56
5	50	13.40

Or $\{C+I(1+L/100)\}\{C+I(1-L/100)\} < (C+I)^2$

Or $C^2 + CI(1-L/100) + CI(1+L/100) + I^2\{1-(L/100)^2\} < (C+I)^2$

Or $C^2 + 2CI + I^2 - I^2(L/100)^2 < (C+I)^2$

Or $(C+I)^2 - I^2(L/100)^2 < (C+I)^2$

Hence proved.

3.4.2 RESTORING THE t VALUE

With a reduction in the I value, the production cycle time t decreases. Such a reduced t might affect the time when the input item is to be procured from the supplier firm. It may also affect the time when the finished assemblies are to be dispatched to the downstream location. In the event when restoring the t value becomes important, a reduction in the holding cost may be explored. Now the t value is 4 months in the basic example. With a reduction in I from ₹500 to ₹400, t value decreases to 3.79. In order to restore the t value, a decreased H, i.e., H_m can be found as follows:

$$t = \sqrt{\frac{2T(C+I)}{(1-D/P)DH}}$$

$$\text{Or } 4 = \sqrt{\frac{2\times36\times(500+400)}{(1-900/1200)\times900H_m}}$$

Or $H_m = ₹18$

In order to generalize,

$L = \%$ reduction in I
$M = \%$ reduction in H

Now,

$$\sqrt{\frac{2T(C+I)}{(1-D/P)DH}} = \sqrt{\frac{2T\left[C+I(1-L/100)\right]}{(1-D/P)DH(1-M/100)}}$$

Or $(C+I)(1-M/100) = C+I(1-L/100)$

$$\text{Or } 1-\frac{M}{100} = \frac{C+I(1-L/100)}{(C+I)}$$

$$\text{Or } \frac{M}{100} = \frac{C+I-C-I(1-L/100)}{(C+I)}$$

$$\text{Or } \frac{M}{100} = \frac{(IL/100)}{(C+I)}$$

$$\text{Or } M = \frac{IL}{(C+I)}$$

In order to know the related variation, Table 3.5 gives the M values corresponding to that of L for the given data.

In a reverse scenario,

L = % reduction in H
M = % reduction in I

Also,

$$\sqrt{\frac{2T(C+I)}{(1-D/P)DH}} = \sqrt{\frac{2T\left[C+I(1-M/100)\right]}{(1-D/P)DH(1-L/100)}}$$

$$\text{Or } (C+I)(1-L/100) = C + I(1-M/100)$$

$$\text{Or } I(1-M/100) = C(1-L/100) + I(1-L/100) - C$$

$$\text{Or } I(1-M/100) = I(1-L/100) - (CL/100)$$

$$\text{Or } 1 - \frac{M}{100} = \frac{I(1-L/100) - (CL/100)}{I}$$

$$\text{Or } \frac{M}{100} = \frac{I - I(1-L/100) + (CL/100)}{I}$$

$$\text{Or } \frac{M}{100} = \frac{(IL/100) + (CL/100)}{I}$$

TABLE 3.5

M (for Decreased H) Corresponding to L (for Decreased I)

S. No.	L	M
1	25	12.5
2	30	15.0
3	40	20.0

$$\text{Or } M = \frac{IL + CL}{I}$$

$$\text{Or } M = \frac{L(I + C)}{I}$$

In order to know the related variation, Table 3.6 gives the M values corresponding to that of L for the given data.

The present values of M are higher than those in the previous situation. It can be verified as follows:

$$\frac{L(I + C)}{I} > \frac{IL}{(C + I)}$$

$$\text{Or } \frac{(I + C)}{I} > \frac{I}{(C + I)}$$

$$\text{Or } Y > \frac{1}{Y}$$

where $Y = \frac{(I + C)}{I}$

And also $Y > 1$

3.4.3 VARIATION IN T

Consider the following example data:

A total planned duration $T = 36$ months
Total demand $D = 900$
Total production capacity $P = 1200$
Facility setup cost $C = ₹500$
Apportioned innovation cost per cycle $I = ₹500$
Inventory holding cost per unit product $H = ₹20$ per month

TABLE 3.6
M (for Decreased I)
Corresponding to L
(for Decreased H)

S. No.	L	M
1	25	50
2	30	60
3	40	80

Now, from Eq. (3.6),

$$t = \sqrt{\frac{2T(C+I)}{(1-D/P)DH}}$$

$$t = \sqrt{\frac{2 \times 36 \times (500+500)}{(1-900/1200) \times 900 \times 20}}$$

Or $t = 4$ months

The total planned duration T is a strategic factor and may be reduced or increased. Table 3.7 shows a reduction in T and the resulting values of t.

For a general approach, let

$L = \%$ reduction in T

A varied t can be given as follows:

$$t_m = \sqrt{\frac{2T(1-L/100)(C+I)}{(1-D/P)DH}}$$

$$\% \text{ reduction in } t = 100\left(\frac{t-t_m}{t}\right)$$

$$= 100\left(1 - \frac{t_m}{t}\right)$$

$$= 100\left\{1 - \sqrt{(1-L/100)}\right\}$$

Table 3.8 shows the % reduction in t for various values of L.

TABLE 3.7

Corresponding t for a Reduced T

S. No.	T	t
1	33	3.83
2	30	3.65
3	27	3.46
4	24	3.27
5	21	3.05

TABLE 3.8

% Reduction in t Corresponding to L

L	1	2	3	4
% reduction in t	0.501	1.005	1.511	2.020

For instance, $t = 3.05$ for a reduced $T = 21$ with reference to Table 3.7. In order to restore the value of t:

$$4 = \sqrt{\frac{2\times 21\times (500 + 500)}{(1 - 900/1200)\times 900 H_m}}$$

Or $H_m = ₹11.67$

If it is possible, the inventory holding cost can be reduced to ₹11.67 for this purpose. However, another option can be a reduction in the production capacity with a revised value P_m for this parameter. Now,

$$4 = \sqrt{\frac{2\times 21\times 1000}{(1 - 900/P_m)\times 900\times 20}}$$

$$\text{Or } 1 - \frac{900}{P_m} = 0.1458333$$

$$\text{Or } P_m = 1053.66$$

For a general approach,

$L = \%$ reduction in T
$M = \%$ reduction in P

Also,

$$\sqrt{\frac{2T(C+I)}{(1-D/P)DH}} = \sqrt{\frac{2T(1-L/100)(C+I)}{\{1-D/P(1-M/100)\}DH}}$$

$$\text{Or } \frac{\{1-D/P(1-M/100)\}}{(1-D/P)} = (1-L/100)$$

$$\text{Or } 1 - \frac{D}{P(1-M/100)} = 1 - \frac{D}{P} - \frac{L}{100}\left(1-\frac{D}{P}\right)$$

$$\text{Or} \quad \frac{D}{P(1-M/100)} = \frac{D}{P} + \frac{L}{100}\left(1-\frac{D}{P}\right)$$

$$\text{Or} \quad \frac{D}{P(1-M/100)} = \frac{D}{P} + \frac{L}{100} - \frac{L}{100}\left(\frac{D}{P}\right)$$

$$\text{Or} \quad \frac{D}{P(1-M/100)} = \frac{100D + LP - LD}{100P}$$

$$\text{Or} \quad \frac{P(1-M/100)}{D} = \frac{100P}{100D + LP - LD}$$

$$\text{Or} \quad 1 - \frac{M}{100} = \frac{100D}{100D + LP - LD}$$

$$\text{Or} \quad \frac{M}{100} = \frac{100D + LP - LD - 100D}{100D + LP - LD}$$

$$\text{Or} \quad M = \frac{100L(P-D)}{100D + L(P-D)}$$

Table 3.9 represents the M values for different L values considering the given data.

In a certain changed environment, the manufacturing organization may plan to increase the horizon strategically. That may happen when a longer planned duration appears to work effectively for a particular innovative product. In case where T increases, the t value also increases. Table 3.10 shows the increased T and the resulting t values.

For a general approach, let

$L = \%$ increase in T

TABLE 3.9

Corresponding M for L (Reduced T)

S. No.	L	M
1	4	1.32
2	8	2.60
3	12	3.85
4	16	5.06

TABLE 3.10

Corresponding *t* for an Increased *T*

S. No.	T	t
1	39	4.16
2	42	4.32
3	45	4.47
4	48	4.62
5	51	4.76

TABLE 3.11

% Increase in *t* Corresponding to *L*

L	1	2	3	4
% increase in *t*	0.499	0.995	1.489	1.980

A varied *t* can be given as follows:

$$t_m = \sqrt{\frac{2T(1+L/100)(C+I)}{(1-D/P)DH}}$$

$$\% \text{ increase in } t = 100\left(\frac{t_m - t}{t}\right)$$

$$= 100\left(\frac{t_m}{t} - 1\right)$$

$$= 100\left\{\sqrt{(1+L/100)} - 1\right\}$$

Table 3.11 shows the % increase in *t* for various values of *L*. This percentage variation is relatively lower in comparison with the previous situation.

For instance, $t = 4.47$ for an increased $T = 45$ with reference to Table 3.10. In order to restore the value of *t*:

$$4 = \sqrt{\frac{2\times45\times(500+I_m)}{(1-900/1200)\times900\times20}}$$

Or $I_m = ₹300$

If it is possible, the innovation cost can be reduced to ₹300 for this purpose. However, another option can be an increase in the production capacity if the operational aspects of the manufacturing firm allow it. With a revised value P_m for this parameter,

$$4 = \sqrt{\frac{2 \times 45 \times 1000}{(1 - 900 / P_m) \times 900 \times 20}}$$

$$\text{Or} \quad 1 - \frac{900}{P_m} = 0.3125$$

$$\text{Or} \quad P_m = 1309.09$$

For a general approach,

$L = \%$ increase in T
$M = \%$ increase in P

Also,

$$\sqrt{\frac{2T(C + I)}{(1 - D / P)DH}} = \sqrt{\frac{2T(1 + L / 100)(C + I)}{\{1 - D / P(1 + M / 100)\}DH}}$$

$$\text{Or} \quad \frac{\{1 - D / P(1 + M / 100)\}}{(1 - D / P)} = (1 + L / 100)$$

$$\text{Or} \quad 1 - \frac{D}{P(1 + M / 100)} = 1 - \frac{D}{P} + \frac{L}{100}\left(1 - \frac{D}{P}\right)$$

$$\text{Or} \quad \frac{D}{P(1 + M / 100)} = \frac{D}{P} - \frac{L}{100}\left(1 - \frac{D}{P}\right)$$

$$\text{Or} \quad \frac{D}{P(1 + M / 100)} = \frac{D}{P} - \frac{L}{100} + \frac{L}{100}\left(\frac{D}{P}\right)$$

$$\text{Or} \quad \frac{D}{P(1 + M / 100)} = \frac{100D - LP + LD}{100P}$$

$$\text{Or} \quad \frac{P(1 + M / 100)}{D} = \frac{100P}{100D - LP + LD}$$

$$\text{Or} \quad 1 + \frac{M}{100} = \frac{100D}{100D - LP + LD}$$

$$\text{Or} \quad \frac{M}{100} = \frac{100D - 100D + LP - LD}{100D - LP + LD}$$

TABLE 3.12

Corresponding *M*

for *L* (Increased *T*)

S. No.	L	M
1	4	1.35
2	8	2.74
3	12	4.17
4	16	5.63

$$\text{Or } M = \frac{100L(P-D)}{100D - L(P-D)}$$

Table 3.12 represents the *M* values for different *L* values considering the given data.

With reference to the increase in *T* and restoring the production cycle time, there can be a joint variation in the innovation cost and the production capacity. Now,

$L = \%$ increase in T

$M = \%$ increase in P

$N = \%$ reduction in I

Also,

$$\sqrt{\frac{2T(C+I)}{(1-D/P)DH}} = \sqrt{\frac{2T(1+L/100)\{C+I(1-N/100)\}}{\{1-D/P(1+M/100)\}DH}}$$

$$\text{Or } \frac{(C+I)}{(1-D/P)} = \frac{(1+L/100)\{C+I(1-N/100)\}}{\{1-D/P(1+M/100)\}}$$

$$\text{Or } (C+I)\{1-D/P(1+M/100)\} = (1-D/P)(1+L/100)\{C+I(1-N/100)\}$$

$$\text{Or } C+I(1-N/100) = \frac{(C+I)\{1-D/P(1+M/100)\}}{(1-D/P)(1+L/100)}$$

$$\text{Or } I(1-N/100) = \frac{(C+I)\{1-D/P(1+M/100)\} - C(1-D/P)(1+L/100)}{(1-D/P)(1+L/100)}$$

$$\text{Or } 1 - \frac{N}{100} = \frac{(C+I)\{1-D/P(1+M/100)\} - C(1-D/P)(1+L/100)}{I(1-D/P)(1+L/100)}$$

$$\text{Or } \frac{N}{100} = \frac{I(1-D/P)(1+L/100) - (C+I)\{1-D/P(1+M/100)\} + C(1-D/P)(1+L/100)}{I(1-D/P)(1+L/100)}$$

Or $\dfrac{N}{100} = \dfrac{(C+I)(1-D/P)(1+L/100)-(C+I)\{1-D/P(1+M/100)\}}{I(1-D/P)(1+L/100)}$

Or $N = 100(C+I)\left[\dfrac{(1-D/P)(1+L/100)-\{1-D/P(1+M/100)\}}{I(1-D/P)(1+L/100)}\right]$

For $L = 25$ and the given data:

$$N = \dfrac{40-4.4M}{(1+M/100)}$$

Combinations of M and N can be obtained as given in Table 3.13 in order to widen the choice.

M is used as input and the N value is obtained. However, in a reverse scenario and with the use of expression,

$$N = \dfrac{40-4.4M}{(1+M/100)}$$

Or $N + \dfrac{NM}{100} = 40-4.4M$

Or $4.4M + \dfrac{NM}{100} = 40-N$

Or $M\{4.4+N/100\} = 40-N$

Or $M = \dfrac{40-N}{\{4.4+N/100\}}$

Now N is used as input and the M value is obtained. Combinations of N and M are given in Table 3.14.

TABLE 3.13

Combinations of M and N (for Higher T)

S. No.	M	N
1	1	35.25
2	2	30.59
3	3	26.02
4	4	21.54
5	5	17.14

TABLE 3.14

Combinations of N and M (for an Increased T)

S. No.	N	M
1	6	7.62
2	12	6.19
3	18	4.80
4	24	3.45
5	30	2.13

Similarly, with reference to a reduction in T, a joint variation in appropriate factors may also be followed. Now,

L = % reduction in T
M = % reduction in P
N = % reduction in H

Also,

$$\sqrt{\frac{2T(C+I)}{(1-D/P)DH}} = \sqrt{\frac{2T(1-L/100)(C+I)}{\{1-D/P(1-M/100)\}DH(1-N/100)}}$$

Or $\dfrac{2T(C+I)}{(1-D/P)DH} = \dfrac{2T(1-L/100)(C+I)}{\{1-D/P(1-M/100)\}DH(1-N/100)}$

Or $\dfrac{1}{(1-D/P)} = \dfrac{(1-L/100)}{\{1-D/P(1-M/100)\}(1-N/100)}$

Or $1 - \dfrac{N}{100} = \dfrac{(1-L/100)(1-D/P)}{\{1-D/P(1-M/100)\}}$

Or $\dfrac{N}{100} = \dfrac{\{1-D/P(1-M/100)\}-(1-L/100)(1-D/P)}{\{1-D/P(1-M/100)\}}$

Or $N = 100\left[\dfrac{\{1-D/P(1-M/100)\}-(1-L/100)(1-D/P)}{\{1-D/P(1-M/100)\}}\right]$

For the given data and $L = 25$,

$$N = \frac{25 - 3.25M}{1 - (M/25)}$$

TABLE 3.15

Combinations of M and N (for a Lower T)

S. No.	M	N
1	1	22.66
2	2	20.11
3	3	17.33
4	4	14.29
5	5	10.94

TABLE 3.16

Combinations of N and M (for a Decreased T)

S. No.	N	M
1	4	6.80
2	8	5.80
3	12	4.69
4	16	3.45
5	20	2.04

Combinations of M and N can be obtained, as shown in Table 3.15, for a wider choice.

M is used as input and the N value is obtained. However, in a reverse scenario and with the use of expression,

$$N = \frac{25 - 3.25M}{1 - (M/25)}$$

$$\text{Or } N - \frac{NM}{25} = 25 - 3.25M$$

$$\text{Or } 3.25M - \frac{NM}{25} = 25 - N$$

$$\text{Or } M(3.25 - N/25) = 25 - N$$

$$\text{Or } M = \frac{25 - N}{3.25 - (N/25)}$$

Now N is used as input and the M value is obtained. Combinations of N and M are given in Table 3.16.

3.5   IMPERFECT BATCH

A production batch might be of two kinds, as shown in Figure 3.7.

In a perfect batch, the whole quantity produced is acceptable from the quality point of view. However, in an imperfect batch, there are certain unacceptable items in a produced quantity. For instance, consider that 900 components are to be produced by a manufacturing firm in 36 months of a planned duration. And it is estimated that the proportion of acceptable components is 0.9 depending on the manufacturing experience. Such experience may relate to the following:

(i)   The specific facility
(ii) Worker's skill
(iii) Raw material quality
(iv) Facility setup parameters

In such a case, the total requirement would be

$$\frac{900}{0.9} = 1000$$

Thus, the total number of components to be produced can be given as follows:

$$\frac{D}{y}$$

where y = proportion of acceptable components in a production batch.

However, since the components are manufactured within the house, the unacceptable number of components can be rejected and there is no need to add those in inventory. Thus, only acceptable components are added to the inventory level.

Consider one production cycle as shown in Figure 3.8. Similar cycle gets repeated for a planned duration on the basis of a certain strategic period.

Now,

Production quantity per cycle = $\dfrac{Dt}{yT}$

Average inventory during the cycle = $\dfrac{V}{2}$

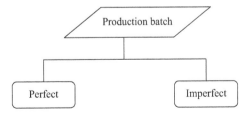

FIGURE 3.7   Perfect and imperfect batch.

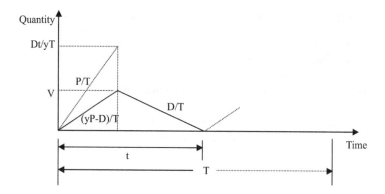

FIGURE 3.8 Production cycle with proportion of acceptable components.

Production inventory buildup happens during the production time only in a cycle. Thus,

$$\frac{V}{(yP-D)/T} = \frac{(Dt/yT)}{(P/T)}$$

$$\text{Or } V = \frac{(1-D/yP)Dt}{T} \tag{3.7}$$

On the basis of average inventory, the holding cost is estimated. Therefore,

Inventory holding cost for the production cycle $= \dfrac{V}{2} \cdot H \cdot t$

With the substitution of Eq. (3.7),

the production inventory holding cost per cycle $= \dfrac{(1-D/yP)Dt}{2T} \cdot H \cdot t$

As there are $\left(\dfrac{T}{t}\right)$ production cycles in the total planned duration,

$$\text{Total production inventory cost} = \frac{(1-D/yP)Dt}{2T} \cdot Ht \cdot \frac{T}{t}$$

$$= \frac{(1-D/yP)DHt}{2} \tag{3.8}$$

After adding the facility setup and innovation cost, the estimated total cost (E) can be expressed as follows:

$$E = \frac{(1-D/yP)DHt}{2} + (C+I)\frac{T}{t} \tag{3.9}$$

In order to get the optimal production cycle time, differentiate with respect to t and equate to zero:

$$\frac{(1-D/yP)DH}{2} - \frac{(C+I)T}{t^2} = 0$$

$$\text{Or } t = \sqrt{\frac{2T(C+I)}{(1-D/yP)DH}} \qquad (3.10)$$

In order to illustrate, consider the following example:

A total planned duration $T = 36$ months
Total demand $D = 900$
Total production capacity $P = 1200$
Proportion of acceptable components in a production batch $y = 0.9$
Facility setup cost $C = ₹500$
Apportioned innovation cost per cycle $I = ₹500$
Inventory holding cost per unit product $H = ₹24$ per month

Now,
From Eq. (3.10),

$$t = \sqrt{\frac{2\times36\times(500+500)}{(1-900/1080)\times900\times24}}$$

Or $t = 4.47$ months
Production cycle time reduces with the increase in y, as shown in Table 3.17.
However, a proportional increase in y may be a suitable factor for a generalized approach in order to know the variation in t.
Let

$L = \%$ increase in y
$t_m = $ value of t with $\%$ variation in y

$$\% \text{ reduction in } t = 100\left(\frac{t-t_m}{t}\right)$$

TABLE 3.17
Variation of t
with Increased y

S. No.	y	t
1	0.9	4.47
2	0.92	4.25
3	0.94	4.06
4	0.96	3.90
5	0.98	3.77

TABLE 3.18

Percentage Reduction in t

S. No.	L	% Reduction in t
1	3	6.57
2	6	11.72
3	9	15.87

TABLE 3.19

Percentage Increase in t

S. No.	L	% Increase in t
1	3	8.76
2	6	21.19
3	9	40.65

$$= 100\left(1 - \frac{t_m}{t}\right)$$

$$= 100\left[1 - \sqrt{\frac{(1 - D/yP)}{1 - D/\{yP(1 + L/100)\}}}\right]$$

Table 3.18 shows the percentage reduction in t with the relevant example data:

Total demand $D = 900$
Total production capacity $P = 1200$
Proportion of acceptable components in a production batch $y = 0.9$

Production cycle time increases with a reduction in y. Now,

$$\% \text{ increase in } t = 100\left(\frac{t_m - t}{t}\right)$$

$$= 100\left(\frac{t_m}{t} - 1\right)$$

$$= 100\left[\sqrt{\frac{(1 - D/yP)}{1 - D/\{yP(1 - L/100)\}}} - 1\right]$$

where $L = \%$ reduction in y

Table 3.19 shows the percentage increase in t with the use of relevant data.

In comparison with the previous situation, the percentage variation in t is much higher.

3.5.1 DECREASE IN T

With the increase in y, the t value decreases. In order to restore the cycle time, a holding cost may be reduced if it is possible. With reference to Table 3.17, when y value increases from 0.90 to 0.94, the t value decreases from 4.47 to 4.06. In order to restore the t value, the revised value of H, i.e., H_m can be obtained as follows:

$$t = \sqrt{\frac{2T(C+I)}{(1-D/yP)DH}}$$

$$\text{Or } 4.47 = \sqrt{\frac{2\times 36\times(500+500)}{\left[1-\left\{900/(0.94\times 1200\right\}\right]900 H_m}}$$

Or $H_m = ₹19.81$

3.5.2 INCREASE IN t

With the reduction in y, the t value increases. In order to restore the cycle time, an innovation cost may be reduced if it is possible. Now,

$L = \%$ reduction in y
$M = \%$ reduction in I

With the use of Eq. (3.10),

$$\sqrt{\frac{2T(C+I)}{(1-D/yP)DH}} = \sqrt{\frac{2T\left\{C+I(1-M/100)\right\}}{\left\{1-D/Py(1-L/100)\right\}DH}}$$

$$\text{Or } \frac{(C+I)}{(1-D/yP)} = \frac{C+I(1-M/100)}{1-D/yP(1-L/100)}$$

$$\text{Or } C+I-(IM/100) = \frac{(C+I)\left\{1-D/yP(1-L/100)\right\}}{(1-D/yP)}$$

$$\text{Or } \frac{IM}{100} = (C+I) - \frac{(C+I)\left\{1-D/yP(1-L/100)\right\}}{(1-D/yP)}$$

$$\text{Or } \frac{IM}{100} = (C+I)\left[1-\frac{\left\{1-D/yP(1-L/100)\right\}}{(1-D/yP)}\right]$$

$$\text{Or} \quad \frac{IM}{100} = (C+I)\left[1 - \frac{\{1 - D/yP(1-L/100)\}}{(1-D/yP)}\right]$$

$$\text{Or} \quad M = 100\left(1 + \frac{C}{I}\right)\left[1 - \frac{\{1 - D/yP(1-L/100)\}}{(1-D/yP)}\right] \qquad (3.11)$$

Consider the relevant example data:

$C = ₹500$
$I = ₹500$
$D = 900$
$P = 1200$
$y = 0.9$

Table 3.20 represents the corresponding values of M for various levels of L when

$$\frac{C}{I} = 1$$

When values of I are more than that of C, it is of interest to consider the reduced values of $\left(\frac{C}{I}\right)$. Table 3.21 gives the interaction of $\left(\frac{C}{I}\right)$ with the values of M for $L = 2$.

In order to generalize, let

$$I = C\left(1 + \frac{N}{100}\right)$$

where the value of I is $N\%$ higher than that of C. And from Eq. (3.11),

$$M = 100\left[1 + \frac{1}{(1+N/100)}\right]\left[1 - \frac{\{1 - D/yP(1-L/100)\}}{(1-D/yP)}\right]$$

For $L = 2$, the obtained values of M corresponding to various levels of N are given in Table 3.22.

TABLE 3.20
Corresponding
M (I) for L (y)

S. No.	L	M
1	1	10.10
2	2	20.41
3	3	30.93

TABLE 3.21

Interaction of $\left(\dfrac{C}{I}\right)$ with the Values of M

S. No.	1	2	3	4	5
$\left(\dfrac{C}{I}\right)$	0.9	0.8	0.7	0.6	0.5
M for $L=2$	19.39	18.37	17.35	16.33	15.31

TABLE 3.22
M Values for Various Levels of N ($I > C$)

S. No.	N	M
1	10	19.48
2	20	18.71
3	30	18.05
4	40	17.49
5	50	17.01

In case where the organization is able to reduce the innovation efforts cost, let

$$I = C\left(1 - \frac{N}{100}\right)$$

where value of I is $N\%$ lower than that of C. And from Eq. (3.11),

$$M = 100\left[1 + \frac{1}{(1-N/100)}\right]\left[1 - \frac{\{1-D/yP(1-L/100)\}}{(1-D/yP)}\right]$$

For $L=2$, the obtained values of M corresponding to various levels of N are provided in Table 3.23.

In comparison with the previous scenario, the present values of M are higher. This is because

$$1 + \frac{1}{(1-N/100)} > 1 + \frac{1}{(1+N/100)}$$

$$\text{Or}\quad \frac{1}{(1-N/100)} > \frac{1}{(1+N/100)}$$

$$\text{Or}\quad (1+N/100) > (1-N/100)$$

TABLE 3.23
**M Values for Various
Levels of N (I < C)**

S. No.	N	M
1	10	21.54
2	20	22.96
3	30	24.78
4	40	27.21
5	50	30.61

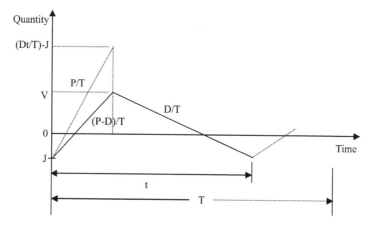

FIGURE 3.9 Permissible backorders.

An extensive analysis is conducted for the perfect and imperfect lot in the context of the production scenario.

3.6 PERMISSIBLE SHORTAGES

When shortages are permitted in the production scenario that are totally backordered, such a situation is shown in Figure 3.9 where

J = maximum permissible shortages that are completely backordered

Production quantity per cycle = $\dfrac{Dt}{T}$

Now,

$$\frac{(V+J)}{(P-D)/T} = \frac{(Dt/T)}{(P/T)}$$

$$\text{Or } V = \frac{Dt(1 - D/P)}{T} - J \tag{3.12}$$

In a cycle, positive inventory exists for the time $= t - \dfrac{JT}{(P-D)} - \dfrac{JT}{D}$

Inventory holding cost in each cycle $= \dfrac{V}{2} \cdot H\left[t - \dfrac{JT}{(P-D)} - \dfrac{JT}{D} \right]$

Total inventory cost $= \dfrac{V}{2} \cdot H\left[t - \dfrac{JT}{(P-D)} - \dfrac{JT}{D} \right] \cdot \dfrac{T}{t}$

Substituting Eq. (3.12),

$$\text{Total inventory cost} = \frac{H}{2}\left[\frac{Dt(1-D/P)}{T} - J \right]\left[t - \frac{JT}{(P-D)} - \frac{JT}{D} \right] \cdot \frac{T}{t}$$

$$= \frac{H}{2}\left[D(1-D/P) - \frac{JT}{t} \right]\left[t - JT\left\{ \frac{1}{D(1-D/P)} \right\} \right]$$

$$= \frac{H}{2}\left[Dt(1-D/P) - JT - JT + \frac{J^2T^2}{tD(1-D/P)} \right]$$

$$= \frac{HDt(1-D/P)}{2} - JTH + \frac{J^2T^2H}{2tD(1-D/P)} \tag{3.13}$$

Time in each cycle when shortages happen $= \dfrac{JT}{(P-D)} + \dfrac{JT}{D}$

$$= \frac{JT}{D(1-D/P)}$$

Total shortage cost $= \dfrac{J}{2} \cdot \dfrac{JT}{D(1-D/P)} \cdot K \cdot \dfrac{T}{t}$

where K = shortage cost per unit for one month
   Also,

$$\text{Total shortage cost} = \frac{J^2T^2K}{2Dt(1-D/P)} \tag{3.14}$$

$$\text{Total setup and innovation cost} = \frac{T}{t}(C + I) \tag{3.15}$$

Adding Eqs. (3.13), (3.14), and (3.15), the total relevant cost can be obtained:

$$E = \frac{HDt(1-D/P)}{2} - JTH + \frac{J^2 T^2 H}{2tD(1-D/P)} + \frac{J^2 T^2 K}{2Dt(1-D/P)} + \frac{T}{t}(C+I)$$

$$\text{Or } E = \frac{HDt(1-D/P)}{2} - JTH + \frac{J^2 T^2 (H+K)}{2Dt(1-D/P)} + \frac{T}{t}(C+I) \qquad (3.16)$$

$\dfrac{\partial E}{\partial J} = 0$ shows:

$$\frac{JT^2(H+K)}{Dt(1-D/P)} - TH = 0$$

$$\text{Or } J = \frac{THDt(1-D/P)}{T^2(H+K)}$$

$$\text{Or } J = \frac{HDt(1-D/P)}{T(H+K)} \qquad (3.17)$$

Using Eq. (3.16), $\dfrac{\partial E}{\partial t} = 0$ shows:

$$\frac{HD(1-D/P)}{2} - \frac{J^2 T^2 (H+K)}{2Dt^2(1-D/P)} - \frac{T(C+I)}{t^2} = 0$$

Substituting Eq. (3.17),

$$\frac{HD(1-D/P)}{2} = \frac{H^2 D(1-D/P)}{2(H+K)} + \frac{T(C+I)}{t^2}$$

$$\text{Or } \frac{HD(1-D/P)}{2}\left[1 - \frac{H}{(H+K)}\right] = \frac{T(C+I)}{t^2}$$

$$\text{Or } \frac{HKD(1-D/P)}{2(H+K)} = \frac{T(C+I)}{t^2}$$

$$\text{Or } t^2 = \frac{2T(H+K)(C+I)}{HDK(1-D/P)}$$

$$\text{Or } t = \sqrt{\frac{2T(H+K)(C+I)}{HDK(1-D/P)}} \qquad (3.18)$$

In order to illustrate, consider the following example:

$T = 36$ months
$D = 900$ number of components
$P = 1200$
$H = ₹20$
$C = ₹400$
$I = ₹400$
$K = ₹80$

The production cycle time can be obtained with the use of Eq. (3.18):

$t = 4$ months

Various aspects pertaining to the production inventory have been discussed and analyzed rigorously, including the permissible shortages that are totally backordered in the system.

4 Multiple Products

Many organizations deal in multiple products. For instance, a finished assembly is composed of many input items and there is requirement of multiple items procurement. Similarly, many items can be manufactured in a product family by an industrial organization. After establishing the need, multiple items procurement as well as manufacture are discussed.

4.1 NEED

Several items are needed for purchase and also for production purposes. Different factors shown in Figure 4.1 explain this requirement for multiple items.

4.1.1 CONSUMPTION PATTERN

One product may be lower in demand in a trading organization and the profitability decreases. In order to compensate, it is appropriate to stock another product that can show a favorable consumption pattern in a time duration. Similarly, few products demonstrate an unfavorable pattern for consumption from the point of view of sales in a trading firm. Thus, multiple items are purchased and can be made available to the end consumers if it is feasible.

4.1.2 STORAGE CAPACITY

Storage capacity refers to the procured as well as produced items, as shown in Figure 4.2.

Multiple items can be procured by a trading firm in order to sell them suitably as time progresses. In the normal course, few products might be available. However, if the storage capacity still allows for some more procured items, a favorable decision in this context may be implemented in order to gain from an overall potentially beneficial scenario. In a manufacturing firm also, many input items are needed for the finished assembly or product at various stages. For the economies of scale, multiple items can be procured if the raw material or input item stores at the factory premises have a proportionate storage capacity.

At the other end of the production line, the finished assemblies or products would be available and need to be stored temporarily before dispatch to the downstream locations/agencies. Multiple products of different specifications can be manufactured if the corresponding storage capacity is available among other desirable supporting factors for this purpose.

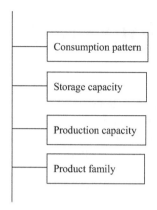

FIGURE 4.1   Contributing factors for multiple items.

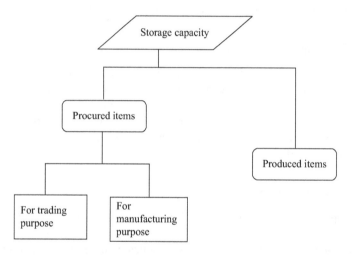

FIGURE 4.2   Items concerning storage.

4.1.3   PRODUCTION CAPACITY

Various factors contribute to an overall assessment of the production capacity, as shown in Figure 4.3.

Machines are set at different speeds, depending on the precise need of a product to be manufactured. Such a predetermined speed can be utilized to estimate the production capacity for certain time duration. Human resources are essentially required in most of the cases for various activities:

  (i) Machine operation
 (ii) Work-in-progress (WIP) handling
(iii) Inspection and testing
(iv) Finished assembly/product handling

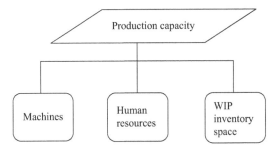

FIGURE 4.3 Components of production capacity assessment.

(v) Packaging
(vi) Loading for dispatch to the downstream locations

Available human resources are an important parameter for an overall assessment of the production capacity associated with certain time duration.

In case where the WIP inventory does not find suitable space between the two machines/workstations, it affects the overall production capacity adversely. Such space should correspond to the desirable capacity of manufacturing multiple products in a given time.

4.1.4 PRODUCT FAMILY

In a product family, there can be multiple items of different specifications and features. Each product in a family may have certain demand in a specified duration. Depending on the production capacity, multiple items can be manufactured in a given time.

Few contributing factors have been discussed in the context of multiple items manufacture and procurement.

4.2 MULTIPLE ITEMS PROCUREMENT

Consider an item i in a set of several items to be procured. For instance, if five items are to be purchased in a given time, then $i = 1, 2, \ldots, 5$. In order to generalize,

$$i = 1, 2, \ldots, n \text{ for } n \text{ number of procured items.}$$

Consider an example where 900 units of demand for an item i is estimated over a period of 3 years or 36 months with almost uniform consumption throughout. In a total period of 36 months, 900 units are to be purchased for an innovative item. Therefore,

The consumption per month $= \dfrac{900}{36}$

In case where a procurement cycle is of 4 months,

the procurement quantity per cycle $= \dfrac{900}{36} \times 4$

In order to generalize, refer to Figure 4.4 where
t = cycle time for procurement in months
T = total planned duration in months

Now,

Procurement quantity per cycle for an item $i = \dfrac{D_i t}{T}$

where D_i = total demand for an item i for the planned duration (T)

Average quantity during the cycle for an item $i = \dfrac{D_i t}{2T}$

Inventory cost during the cycle for an item $i = \dfrac{D_i t}{2T} \cdot H_i \cdot t$

where H_i = inventory carrying cost for an item i per unit for one month

Total inventory cost for the planned duration (T) for an item $i = \dfrac{D_i t}{2T} \cdot H_i t \cdot \dfrac{T}{t}$

$$= \dfrac{D_i t H_i}{2} \tag{4.1}$$

$$\text{Total ordering cost for an item } i = \dfrac{T}{t} \cdot C_i \tag{4.2}$$

where C_i = fixed ordering cost for an item i

$$\text{Total innovation cost for procurement for an item } i = \dfrac{T}{t} \cdot I_i \tag{4.3}$$

where I_i = apportioned innovation cost for one cycle for an item i
    Adding Eqs. (4.1), (4.2), and (4.3),

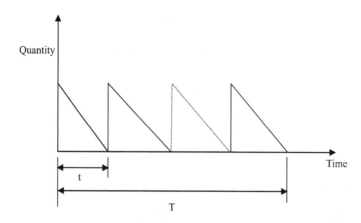

FIGURE 4.4   Purchase pattern for an item i over a total planned duration.

Estimated total cost for an item $i = \dfrac{D_i t H_i}{2} + \dfrac{T}{t} \cdot (C_i + I_i)$

For the multiple items procurement, i.e., for $i = 1, 2, \ldots, n$ items, the total cost for the planned duration is as follows:

$$E = \frac{t}{2} \sum_{i=1}^{n} D_i H_i + \frac{T}{t} \sum_{i=1}^{n} (C_i + I_i) \tag{4.4}$$

Optimal value of t can be obtained by differentiating Eq. (4.4) with respect to t and equating to zero:

$$\frac{dE}{dt} = \frac{1}{2} \sum_{i=1}^{n} D_i H_i - \frac{T}{t^2} \sum_{i=1}^{n} (C_i + I_i) = 0$$

$$\text{Or } t^2 = \frac{2T \sum\limits_{i=1}^{n} (C_i + I_i)}{\sum\limits_{i=1}^{n} D_i H_i}$$

$$\text{Or } t = \sqrt{\frac{2T \sum\limits_{i=1}^{n} (C_i + I_i)}{\sum\limits_{i=1}^{n} D_i H_i}} \tag{4.5}$$

4.2.1 ILLUSTRATIVE EXAMPLE

In order to illustrate, consider the three products with a total planned duration:

$$T = 36 \text{ months}$$

Table 4.1 shows the parameters for these products.
From Eq. (4.5), the procurement cycle time can be obtained as follows:

$$t = \sqrt{\frac{2 \times 36 \times \left[700 + 1200 + 5300 \right]}{57600}}$$

Or $t = 3$ months

4.2.2 INNOVATIVE PRODUCT ENTRY

A firm is dealing with a certain number of innovative items. There is a proposal to add the next product, but this influences the current scenario. In order to illustrate, consider the two items with a total planned duration:

$$T = 36 \text{ months}$$

TABLE 4.1

Parameters for the Products

Parameters	$i = 1$	$i = 2$	$i = 3$
Fixed ordering cost C_i	₹300	₹500	₹800
Apportioned innovation cost I_i	₹400	₹700	₹4500
Total demand D_i	1800	2160	1440
Inventory carrying cost H_i	₹8	₹10	₹15

TABLE 4.2

Parameters for Both Items

Parameters	$i = 1$	$i = 2$
Fixed ordering cost C_i	₹150	₹250
Apportioned innovation cost I_i	₹150	₹250
Total demand D_i	360	540
Inventory carrying cost H_i	₹4	₹4

Table 4.2 shows the parameters for both the items.

From Eq. (4.5), the procurement cycle time can be obtained:

$$t = \sqrt{\frac{2 \times 36 \times 800}{(1440 + 2160)}}$$

Or $t = 4$ months

The third product enters with the parameter values as follows:

Fixed ordering cost $C_3 = ₹400$
Apportioned innovation cost $I_3 = ₹1230$
Total demand $D_3 = 1980$
Inventory carrying cost $H_3 = ₹8$

Now, for these three items, from Eq. (4.5), the procurement cycle time can be obtained:

$$t = \sqrt{\frac{2 \times 36 \times [800 + 1630]}{(3600 + 15840)}}$$

Or $t = 3$ months

However, this reduced procurement cycle time from 4 to 3 months may not be suitable for other companies in the chain such as the manufacturer. In order to restore the cycle time,

$$4 = \sqrt{\frac{2 \times 36 \times 2430}{\sum\limits_{i=1}^{i=3} D_i H_i}}$$

$$\text{Or } \sum_{i=1}^{i=3} D_i H_i = 10935$$

In case where it is possible to reduce the carrying cost of the entered item, this aim of restoring the cycle time can be achieved. Now,

$$D_3 H_3 = 10935 - 3600 = 7335$$

Or $H_3 = ₹3.70$

In order to generalize,

$$t = \sqrt{\frac{2T \sum\limits_{i=1}^{n} (C_i + I_i)}{\sum\limits_{\substack{i=1 \\ i \neq j}}^{n} (D_i H_i) + D_j H_j}}$$

$$\text{Or } t^2 = \frac{2T \sum\limits_{i=1}^{n} (C_i + I_i)}{\sum\limits_{\substack{i=1 \\ i \neq j}}^{n} (D_i H_i) + D_j H_j}$$

$$\text{Or } D_j H_j + \sum_{\substack{i=1 \\ i \neq j}}^{n} (D_i H_i) = \frac{2T \sum\limits_{i=1}^{n} (C_i + I_i)}{t^2}$$

$$\text{Or } D_j H_j = \frac{2T \sum\limits_{i=1}^{n} (C_i + I_i) - t^2 \sum\limits_{\substack{i=1 \\ i \neq j}}^{n} (D_i H_i)}{t^2}$$

$$\text{Or } H_j = \frac{2T \sum\limits_{i=1}^{n} (C_i + I_i) - t^2 \sum\limits_{\substack{i=1 \\ i \neq j}}^{n} (D_i H_i)}{D_j t^2}$$

In another case where the firm does not have any idea about the innovation cost of an item and that is under consideration for entry, an idea can be generated for the suitable innovation cost in the context of the present issue. It may be implemented in case where it is feasible. Now,

$$t = \sqrt{\frac{2T\left\{\sum_{i=1}^{n}(C_i) + \sum_{\substack{i=1 \\ i \neq j}}^{n}(I_i) + I_j\right\}}{\sum_{i=1}^{n}(D_iH_i)}}$$

Or $t^2 \sum_{i=1}^{n}(D_iH_i) = 2T\left\{\sum_{i=1}^{n}(C_i) + \sum_{\substack{i=1 \\ i \neq j}}^{n}(I_i)\right\} + 2TI_j$

Or $2TI_j = t^2 \sum_{i=1}^{n}(D_iH_i) - 2T\left\{\sum_{i=1}^{n}(C_i) + \sum_{\substack{i=1 \\ i \neq j}}^{n}(I_i)\right\}$

Or $I_j = \dfrac{t^2 \sum_{i=1}^{n}(D_iH_i) - 2T\left\{\sum_{i=1}^{n}(C_i) + \sum_{\substack{i=1 \\ i \neq j}}^{n}(I_i)\right\}}{2T}$

4.2.3   INNOVATIVE PRODUCT EXIT

Consider the three items as shown in Table 4.3 for their parameters.

For these three items and the total planned duration of 36 months, from Eq. (4.5), the procurement cycle time can be obtained:

$$t = \sqrt{\frac{2 \times 36 \times 2430}{19440}}$$

Or $t = 3$ months

Because of some strategic reasons, the third item needs to be discarded. When this product exits, the scenario will return to

$$t = 4 \text{ months}$$

However, in case where this changed t value is not suitable, the innovation cost may be focused on. Since now the first two items remain, assume that the second item, i.e., $j = 2$, is chosen for the innovation cost reduction if it is feasible. In order to restore the procurement cycle time,

TABLE 4.3
Parameters for the Three Items

Parameters	$i = 1$	$i = 2$	$i = 3$
Fixed ordering cost C_i	₹150	₹250	₹400
Apportioned innovation cost I_i	₹150	₹250	₹1230
Total demand D_i	360	540	1980
Inventory carrying cost H_i	₹4	₹4	₹8

$$3 = \sqrt{\frac{2 \times 36 \times (150 + 150 + 250 + I_j)}{3600}}$$

$$\text{Or} \quad \frac{9 \times 3600}{72} = 550 + I_j$$

$$\text{Or} \quad 450 = 550 + I_j$$

As it is not feasible,

$$3 = \sqrt{\frac{2 \times 36 \times \left(150 + 250 + \sum_{i=1}^{2} I_i\right)}{3600}}$$

$$\text{Or} \quad \frac{9 \times 3600}{72} = 400 + \sum_{i=1}^{2} I_i$$

$$\text{Or} \quad \sum_{i=1}^{2} I_i = 450 - 400 = 50$$

Certain insights can be obtained for the potential reduction in the joint innovation cost if that is feasible.

In order to generalize,

$$t = \sqrt{\frac{2T\left\{\sum_{i=1}^{n}(C_i) + \sum_{i=1}^{n}(I_i)\right\}}{\sum_{i=1}^{n}(D_i H_i)}}$$

$$\text{Or} \quad \frac{t^2}{2T}\sum_{i=1}^{n}(D_i H_i) = \sum_{i=1}^{n}(C_i) + \sum_{i=1}^{n}(I_i)$$

$$\text{Or } \sum_{i=1}^{n}(I_i) = \frac{t^2\sum_{i=1}^{n}(D_iH_i) - 2T\sum_{i=1}^{n}(C_i)}{2T}$$

In order to explore the feasibility of one innovative product j,

$$I_j = \frac{t^2\sum_{i=1}^{n}(D_iH_i) - 2T\left\{\sum_{i=1}^{n}(C_i) + \sum_{\substack{i=1\\i\neq j}}^{n}(I_i)\right\}}{2T}$$

4.2.4 VARIATION IN THE PLANNING HORIZON

Depending on a specific environment of innovative products, the planning horizon should be strategically chosen. It might vary from few months to few years. Therefore, a variation in the total planned duration is analyzed along with its impact on the procurement cycle time. First, a reduction in the planning horizon and thereafter an increase in this duration is discussed.

In case where the total planned duration (T) reduces, the procurement cycle time (t) also reduces. With reference to Table 4.2, the parameters for the two products are available. As the total planned duration, $T = 36$ months, from Eq. (4.5),

$$t = \sqrt{\frac{2\times 36\times 800}{(1440 + 2160)}}$$

Or $t = 4$ months

Table 4.4 shows the variation in the planning horizon, i.e., T is reduced from 36 months. And corresponding t values are also given in months.

For a generalization, let

$L = \%$ reduction in T
$M = \%$ reduction in t

TABLE 4.4
Reduction in the Planning Horizon

S. No.	T	t
1	36	4.00
2	32	3.77
3	28	3.53
4	24	3.27
5	20	2.98

For the reduced T, from Eq. (4.5),

$$t\left(1-\frac{M}{100}\right) = \sqrt{\frac{2T(1-L/100)\left\{\displaystyle\sum_{i=1}^{n}(C_i+I_i)\right\}}{\displaystyle\sum_{i=1}^{n}D_iH_i}}$$

Or $\quad 1-\dfrac{M}{100} = \dfrac{1}{t}\sqrt{\dfrac{2T(1-L/100)\left\{\displaystyle\sum_{i=1}^{n}(C_i+I_i)\right\}}{\displaystyle\sum_{i=1}^{n}D_iH_i}}$

Or $\quad \dfrac{M}{100} = 1-\left(\dfrac{1}{t}\right)\sqrt{\dfrac{2T(1-L/100)\left\{\displaystyle\sum_{i=1}^{n}(C_i+I_i)\right\}}{\displaystyle\sum_{i=1}^{n}D_iH_i}}$

Or $\quad M = 100\left[1-\left(\dfrac{1}{t}\right)\sqrt{\dfrac{2T(1-L/100)\left\{\displaystyle\sum_{i=1}^{n}(C_i+I_i)\right\}}{\displaystyle\sum_{i=1}^{n}D_iH_i}}\right]$ (4.6)

From Eq. (4.5),

$$t = \sqrt{\frac{2T\displaystyle\sum_{i=1}^{n}(C_i+I_i)}{\displaystyle\sum_{i=1}^{n}D_iH_i}}$$

Or $\quad t\sqrt{(1-L/100)} = \sqrt{\dfrac{2T(1-L/100)\displaystyle\sum_{i=1}^{n}(C_i+I_i)}{\displaystyle\sum_{i=1}^{n}D_iH_i}}$

Substituting in Eq. (4.6),

$$M = 100\left[1-\left(\frac{1}{t}\right)t\sqrt{(1-L/100)}\right]$$

$$\text{Or } M = 100\left[1 - \sqrt{(1 - L/100)}\right]$$

Table 4.5 represents the corresponding values of M.

When the T value increases, the procurement cycle time also increases. Table 4.6 shows the variation in the planning horizon, i.e., T is increased from 36 months. And corresponding t values are also given in months.

For a generalization, let

$L = \%$ increase in T
$M = \%$ increase in t

For the increased T, Eq. (4.5) can be used to obtain the varied t:

$$t\left(1 + \frac{M}{100}\right) = \sqrt{\frac{2T(1 + L/100)\left\{\sum_{i=1}^{n}(C_i + I_i)\right\}}{\sum_{i=1}^{n} D_i H_i}}$$

TABLE 4.5
Corresponding Values of M (for a Reduced T)

S. No.	L	M
1	4	2.02
2	8	4.08
3	12	6.19
4	16	8.35

TABLE 4.6
Increase in the Planning Horizon

S. No.	T	t
1	36	4.00
2	40	4.22
3	44	4.42
4	48	4.62
5	52	4.81

TABLE 4.7

Corresponding Values of
M (for an Increased T)

S. No.	L	M
1	4	1.98
2	8	3.92
3	12	5.83
4	16	7.70

$$\text{Or } t\left(1+\frac{M}{100}\right) = t\sqrt{(1+L/100)}$$

$$\text{Or } 1+\frac{M}{100} = \sqrt{(1+L/100)}$$

$$\text{Or } \frac{M}{100} = \sqrt{(1+L/100)} - 1$$

$$\text{Or } M = 100\left[\sqrt{(1+L/100)} - 1\right]$$

Table 4.7 represents the corresponding values of M.

Comparing with the previous situation, the M values are lower. This can be proved as follows:

$$\sqrt{(1+L/100)} - 1 < 1 - \sqrt{(1-L/100)}$$

$$\text{Or } \sqrt{(1+L/100)} + \sqrt{(1-L/100)} < 2$$

$$\text{Or } 1+(L/100)+1-(L/100)+2\sqrt{1-(L/100)^2} < 4$$

$$\text{Or } 1+\sqrt{1-(L/100)^2} < 2$$

$$\text{Or } \sqrt{1-(L/100)^2} < 1$$

L can be less than 100 for all practical purposes, and hence it is verified.

4.3 MULTIPLE ITEMS MANUFACTURE

Consider an item i in a set of several items to be manufactured. For instance, if five items are to be produced in a given time, then $i = 1, 2, ..., 5$. In order to generalize,

$i = 1, 2, ..., n$ for n number of produced items.

Consider an example where 900 units of demand for an item i is estimated over a period of 3 years, or 36 months, with almost uniform consumption throughout. In a total period of 36 months, 900 units are to be manufactured for an innovative item. Therefore,

The demand per month $= \dfrac{900}{36}$

In case where a manufacturing cycle is of 4 months,

the produced quantity per cycle $= \dfrac{900}{36} \times 4$

Consider one manufacturing cycle for an item i, as shown in Figure 4.5. Similar cycle gets repeated for a planned duration on the basis of a certain strategic period.

In order to generalize, refer to Figure 4.5 where

t = production cycle time in months
T = total planned duration in months

Now,

Production quantity per cycle for an item $i = \dfrac{D_i t}{T}$

where D_i = total demand of an item i for the planned duration (T)

Average inventory during the cycle for an item $i = \dfrac{V_i}{2}$

where V_i = maximum production inventory during the cycle for an item i.

Production inventory buildup for an item i happens during the production time only in a cycle. Thus,

$$\frac{V_i}{(P_i - D_i)/T} = \frac{(D_i t / T)}{(P_i / T)}$$

where P_i = total production capacity for an item i for the total duration T

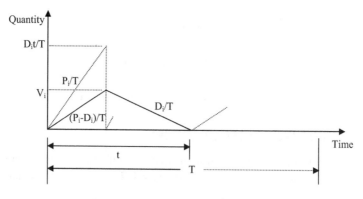

FIGURE 4.5 Manufacturing cycle for an item i and continuation.

$$\text{Or } V_i = \frac{(1-D_i/P_i)D_i t}{T} \tag{4.7}$$

On the basis of average inventory, the holding cost is estimated. Therefore,

$$\text{Inventory holding cost for the production cycle} = \frac{V_i}{2} \cdot H_i \cdot t$$

where H_i = inventory holding cost per item for an item i in ₹/month

With the substitution of Eq. (4.7),

$$\text{the production inventory holding cost for an item } i \text{ per cycle} = \frac{(1-D_i/P_i)D_i t}{2T} \cdot H_i \cdot t$$

As there are $\left(\dfrac{T}{t}\right)$ production cycles in the total planned duration,

$$\text{total production inventory cost for an item } i = \frac{(1-D_i/P_i)D_i t}{2T} \cdot H_i t \cdot \frac{T}{t}$$

$$= \frac{(1-D_i/P_i)D_i H_i t}{2} \tag{4.8}$$

$$\text{Total setup and innovation cost for an item } i = (C_i + I_i) \cdot \frac{T}{t} \tag{4.9}$$

where C_i = fixed setup cost per facility setup for an item i

and I_i = apportioned innovation cost per production cycle for an item i

Estimated total cost for an item i can be obtained by adding Eqs. (4.8) and (4.9) as given below:

$$\frac{(1-D_i/P_i)D_i H_i t}{2} + (C_i + I_i)\frac{T}{t}$$

For the multiple items manufacture, i.e., for $i = 1, 2, ..., n$ items, the total cost for the planned duration is as follows:

$$E = \frac{t}{2}\sum_{i=1}^{n}(1-D_i/P_i)D_i H_i + \frac{T}{t}\sum_{i=1}^{n}(C_i + I_i) \tag{4.10}$$

Optimal value of t can be obtained by differentiating Eq. (4.10) with respect to t and equating to zero:

$$\frac{dE}{dt} = \frac{1}{2}\sum_{i=1}^{n}(1-D_i/P_i)D_i H_i - \frac{T}{t^2}\sum_{i=1}^{n}(C_i + I_i) = 0$$

$$\text{Or } t^2 = \frac{2T\sum_{i=1}^{n}(C_i + I_i)}{\sum_{i=1}^{n}(1-D_i/P_i)D_i H_i}$$

$$\text{Or } t = \sqrt{\frac{2T \sum_{i=1}^{n}(C_i + I_i)}{\sum_{i=1}^{n}(1 - D_i / P_i)D_i H_i}} \qquad (4.11)$$

4.3.1 EXAMPLE

In order to illustrate, consider the two items with a total planned duration:

$$T = 36 \text{ months}$$

Table 4.8 shows the parameters for both the items.
From Eq. (4.11), the production cycle time can be obtained:

$$t = \sqrt{\frac{2 \times 36 \times 5472}{(4032 + 6912)}}$$

Or $t = 6$ months

4.3.2 INCLUSION OF INNOVATIVE ITEM

With reference to the previous example, a manufacturing firm deals with the two products. However, now it wishes to include a third item because its apportioned innovation cost appears to be relatively lower, i.e., ₹500. It is estimated that the facility setup cost would be ₹492 approximately. Demand for the total planned horizon of 36 months is 630. Production capacity for this horizon is available as 3150 units. Inventory holding cost for this potentially included item is ₹36 per month.

After the inclusion of this item, the parameters are now shown in Table 4.9.
From Eq. (4.11), the production cycle time can be obtained:

$$t = \sqrt{\frac{2 \times 36 \times (5472 + 992)}{(4032 + 6912 + 18144)}}$$

Or $t = 4$ months

TABLE 4.8
Parameters for Both Items

Parameters	$i = 1$	$i = 2$
Fixed setup cost C_i	₹1570	₹1202
Apportioned innovation cost I_i	₹1500	₹1200
Total demand D_i	360	540
Inventory holding cost H_i	₹16	₹16
Total production capacity P_i	1200	2700

TABLE 4.9

Parameters for the Three Produced Items

Parameters	$i = 1$	$i = 2$	$i = 3$
Fixed setup cost C_i	₹1570	₹1202	₹492
Apportioned innovation cost I_i	₹1500	₹1200	₹500
Total demand D_i	360	540	630
Inventory holding cost H_i	₹16	₹16	₹36
Total production capacity P_i	1200	2700	3150

However, if it is not suitable for either the firm or other companies in the chain, the t value might be restored also.

$$6 = \sqrt{\frac{2 \times 36 \times 6464}{\sum_{i=1}^{i=3}(1 - D_i / P_i)D_iH_i}}$$

$$\text{Or } \sum_{i=1}^{i=3}(1 - D_i / P_i)D_iH_i = 12928$$

In case where it is possible to reduce the carrying cost of the entered item, this aim of restoring the cycle time can be achieved. Now,

$$(1 - D_3 / P_3)D_3H_3 = 12928 - 10944 = 1984$$

Or $H_3 = ₹3.94$

In order to generalize,

$$t = \sqrt{\frac{2T\sum_{i=1}^{n}(C_i + I_i)}{\sum_{i=1}^{n}(1 - D_i / P_i)D_iH_i}}$$

$$\text{Or } \sum_{i=1}^{n}(1 - D_i / P_i)D_iH_i = \frac{2T\sum_{i=1}^{n}(C_i + I_i)}{t^2}$$

$$\text{Or } (1 - D_j / P_j)D_jH_j = \frac{2T\sum_{i=1}^{n}(C_i + I_i)}{t^2} - \sum_{\substack{i=1 \\ i \neq j}}^{n}(1 - D_i / P_i)D_iH_i$$

$$\text{Or } (1-D_j/P_j)D_jH_j = \frac{2T\sum_{i=1}^{n}(C_i+I_i)-t^2\sum_{\substack{i=1\\i\neq j}}^{n}(1-D_i/P_i)D_iH_i}{t^2}$$

$$\text{Or } H_j = \frac{2T\sum_{i=1}^{n}(C_i+I_i)-t^2\sum_{\substack{i=1\\i\neq j}}^{n}(1-D_i/P_i)D_iH_i}{t^2(1-D_j/P_j)D_j}$$

In another case where the firm does not have any idea about the innovation cost of an item and that is under consideration for entry, an idea can be generated for the suitable innovation cost in the context of the present issue. It may be implemented in case where it is feasible. Now,

$$t = \sqrt{\frac{2T\left\{\sum_{i=1}^{n}(C_i)+\sum_{\substack{i=1\\i\neq j}}^{n}(I_i)+I_j\right\}}{\sum_{i=1}^{n}(1-D_i/P_i)D_iH_i}}$$

$$\text{Or } t^2\sum_{i=1}^{n}(1-D_i/P_i)D_iH_i = 2T\left\{\sum_{i=1}^{n}(C_i)+\sum_{\substack{i=1\\i\neq j}}^{n}(I_i)+I_j\right\}$$

$$\text{Or } \sum_{i=1}^{n}(C_i)+\sum_{\substack{i=1\\i\neq j}}^{n}(I_i)+I_j = \frac{t^2\sum_{i=1}^{n}(1-D_i/P_i)D_iH_i}{2T}$$

$$\text{Or } I_j = \frac{t^2\sum_{i=1}^{n}(1-D_i/P_i)D_iH_i}{2T} - \left\{\sum_{i=1}^{n}(C_i)+\sum_{\substack{i=1\\i\neq j}}^{n}(I_i)\right\}$$

$$\text{Or } I_j = \frac{t^2\sum_{i=1}^{n}(1-D_i/P_i)D_iH_i-2T\left\{\sum_{i=1}^{n}(C_i)+\sum_{\substack{i=1\\i\neq j}}^{n}(I_i)\right\}}{2T}$$

4.3.3 EXCLUSION OF INNOVATIVE ITEM

Consider the existing scenario of the three items as represented in Table 4.9. For instance, if the third item is discarded, i.e., the manufacturing company is not interested to produce this item. After exclusion, the situation returns to the two items manufacture, and the production cycle time increases from 4 to 6 months. In order to restore this cycle time,

$$4 = \sqrt{\frac{2 \times 36 \sum_{i=1}^{2}(C_i + I_i)}{10,944}}$$

$$\text{Or } 16 \times 10,944 = 72 \sum_{i=1}^{2}(C_i + I_i)$$

$$\text{Or } \sum_{i=1}^{2}(C_i + I_i) = 2432$$

However, currently, $\sum_{i=1}^{2}(C_i + I_i) = 5472$, therefore, a reduction of ₹3040 may be needed jointly.

In order to generalize,

$$t = \sqrt{\frac{2T \sum_{i=1}^{n}(C_i + I_i)}{\sum_{i=1}^{n}(1 - D_i / P_i)D_i H_i}}$$

$$\text{Or } t^2 \sum_{i=1}^{n}(1 - D_i / P_i)D_i H_i = 2T \sum_{i=1}^{n}(C_i + I_i)$$

$$\text{Or } \sum_{i=1}^{n}(C_i + I_i) = \frac{t^2}{2T} \sum_{i=1}^{n}(1 - D_i / P_i)D_i H_i$$

The above expression can be used for implementation. However, depending on the requirement, joint innovation cost may also be analyzed as follows:

$$\sum_{i=1}^{n}(I_i) = \frac{t^2 \left\{ \sum_{i=1}^{n}(1 - D_i / P_i)D_i H_i \right\} - 2T \sum_{i=1}^{n}(C_i)}{2T}$$

In order to explore the feasibility of one innovative product j,

$$I_j = \frac{t^2 \left\{ \sum_{i=1}^{n} (1 - D_i / P_i) D_i H_i \right\} - 2T \left\{ \sum_{i=1}^{n} (C_i) + \sum_{\substack{i=1 \\ i \neq j}}^{n} (I_i) \right\}}{2T}$$

4.3.4 CHANGE IN THE PLANNED DURATION

Consider the two items manufacture with a total planned duration:

$$T = 36 \text{ months}$$

With reference to Table 4.8, the parameters for both the items are given. Now, from Eq. (4.11), the production cycle time can be obtained:

$$t = \sqrt{\frac{2 \times 36 \times 5472}{(4032 + 6912)}}$$

Or $t = 6$ months

The total planned duration, T is a strategic parameter and it can vary in practice. This variation affects the production cycle time. When the T value increases, the production cycle time t also increases. Table 4.10 shows the variation in the planned duration, i.e., T is increased from 36 months. And the corresponding t values are also given in months.

When the T value decreases, the production cycle time t also decreases. Table 4.11 shows the variation in the planned duration, i.e., T is reduced from 36 months. And the corresponding t values are also given in months.

If a manufacturing organization wishes to alter the total planned duration in the context of multiple items manufacture, its impact on the production cycle time should be observed and analyzed thoroughly before implementation.

TABLE 4.10
Increase in the Planned Duration

S. No.	T	t
1	36	6.00
2	40	6.32
3	44	6.63
4	48	6.93
5	52	7.21

TABLE 4.11
Reduction in the
Planned Duration

S. No.	T	t
1	36	6.00
2	32	5.66
3	28	5.29
4	24	4.90
5	20	4.47

4.4 IMPERFECT ITEMS

Imperfect or defective items are available in a production or purchase lot in many situations. In the context of multiple products, these are incorporated in both scenarios: production and purchase.

4.4.1 PURCHASE SCENARIO

When multiple kinds of components are purchased and a realistic situation of imperfect items is included, then the following parameter can be added in the analysis:

y_i = proportion of acceptable components in a batch for an item i

And the total number of components ordered for purchase for an item i can be given:

$$\frac{D_i}{y_i}$$

Now, two situations can arise. In the first case, the procured quantity in each cycle is adjusted by this proportion as shown in Figure 4.6; however, the unacceptable components are not added to the inventory.

This can happen when the unacceptable components are identified as soon as these are received, and are not added to the inventory. Thus, the inventory carrying costs are not affected. However, the purchased quantity for an item i in each cycle needs to be adjusted with certain proportion and that may impact the purchase cost.

In the second situation, it may not be possible to identify the unacceptable components instantaneously on receipt, and these are added to the inventory level. This affects the inventory costs as well as the procurement cycle time. Now,

Procurement quantity per cycle for an item $i = \dfrac{D_i t}{y_i T}$

Average quantity during the cycle for an item $i = \dfrac{D_i t}{2 y_i T}$

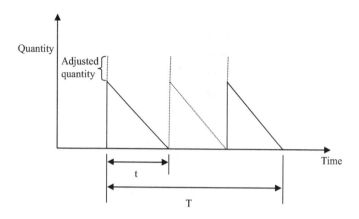

FIGURE 4.6 Imperfect components with no addition to inventory for an item i.

Inventory cost during the cycle for an item $i = \dfrac{D_i t}{2 y_i T} \cdot H_i \cdot t$

Total inventory cost for the planned duration (T) for an item $i = \dfrac{D_i t}{2 y_i T} \cdot H_i t \cdot \dfrac{T}{t}$

$$= \dfrac{D_i t H_i}{2 y_i} \tag{4.12}$$

$$\text{Total ordering cost for an item } i = \dfrac{T}{t} \cdot C_i \tag{4.13}$$

$$\text{Total innovation cost for an item } i = \dfrac{T}{t} \cdot I_i \tag{4.14}$$

Adding Eqs. (4.12), (4.13), and (4.14):

$$\text{Estimated total cost for an item } i = \dfrac{D_i t H_i}{2 y_i} + \dfrac{T}{t} \cdot (C_i + I_i)$$

For the multiple items procurement, i.e., for $i = 1, 2, \ldots, n$ items, the total cost for the planned duration is as follows:

$$E = \dfrac{t}{2} \sum_{i=1}^{n} (D_i H_i / y_i) + \dfrac{T}{t} \sum_{i=1}^{n} (C_i + I_i) \tag{4.15}$$

Optimal value of t can be obtained by differentiating Eq. (4.15) with respect to t and equating to zero:

$$\dfrac{dE}{dt} = \dfrac{1}{2} \sum_{i=1}^{n} (D_i H_i / y_i) - \dfrac{T}{t^2} \sum_{i=1}^{n} (C_i + I_i) = 0$$

$$\text{Or } t^2 = \frac{2T \sum_{i=1}^{n} (C_i + I_i)}{\sum_{i=1}^{n} (D_i H_i / y_i)}$$

$$\text{Or } t = \sqrt{\frac{2T \sum_{i=1}^{n} (C_i + I_i)}{\sum_{i=1}^{n} (D_i H_i / y_i)}} \qquad (4.16)$$

When defective items are added to the inventories, the purchase cycle time needs to be revisited in the context of the imperfect batch. In order to illustrate, consider the three products with a total planned duration:

$$T = 36 \text{ months}$$

Table 4.12 shows the parameters for these products, including the proportion of acceptable components in the batch.

From Eq. (4.16), the purchase cycle time can be obtained:

$$t = \sqrt{\frac{2 \times 36 \times 7200}{60777.658}}$$

Or $t = 2.92$ months

4.4.2 PRODUCTION SCENARIO

In a perfect batch, the whole quantity produced is acceptable from the quality point of view. However, in an imperfect batch, there are certain unacceptable items in a produced quantity. Since the components are manufactured within the house, the unacceptable number of components can be rejected and there is no

TABLE 4.12
Parameters for the Items

Parameters	$i = 1$	$i = 2$	$i = 3$
Fixed ordering cost C_i	₹300	₹500	₹800
Apportioned innovation cost I_i	₹400	₹700	₹4500
Total demand D_i	1800	2160	1440
Inventory carrying cost H_i	₹8	₹10	₹15
Proportion of acceptable components y_i	0.90	0.95	0.98

need to add those in inventory. Thus, only acceptable components are added to the inventory level.

The total number of components to be produced for an item i can be given:

$$\frac{D_i}{y_i}$$

where y_i = proportion of acceptable components in a production batch for an item i.

Consider one production cycle for an item i, as shown in Figure 4.7. Similar cycle gets repeated for a planned duration on the basis of a certain strategic period.

Now,

Production quantity per cycle for an item $i = \dfrac{D_i t}{y_i T}$

Average inventory during the cycle for an item $i = \dfrac{V_i}{2}$

Production inventory buildup for an item i happens during the production time only in a cycle. Thus,

$$\frac{V_i}{(y_i P_i - D_i)/T} = \frac{(D_i t / y_i T)}{(P_i / T)}$$

$$\text{Or } V_i = \frac{(1 - D_i / y_i P_i)D_i t}{T} \qquad (4.17)$$

On the basis of average inventory, the holding cost is estimated. Therefore,

Inventory holding cost for the production cycle $= \dfrac{V_i}{2} \cdot H_i \cdot t$

With the substitution of Eq. (4.17),

The production inventory holding cost for an item i per cycle $= \dfrac{(1 - D_i / y_i P_i)D_i t}{2T} \cdot H_i \cdot t$

As there are $\left(\dfrac{T}{t}\right)$ production cycles in the total planned duration,

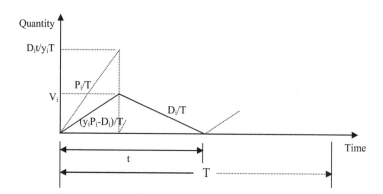

FIGURE 4.7   Production cycle for an item i and continuation.

Total production inventory cost for an item $i = \dfrac{(1 - D_i / y_i P_i) D_i t}{2T} \cdot H_i t \cdot \dfrac{T}{t}$

$$= \frac{(1 - D_i / y_i P_i) D_i H_i t}{2} \qquad (4.18)$$

Total setup and innovation cost for an item $i = (C_i + I_i) \cdot \dfrac{T}{t} \qquad (4.19)$

Estimated total cost for an item i can be obtained by adding Eqs. (4.18) and (4.19) as given below:

$$\frac{(1 - D_i / y_i P_i) D_i H_i t}{2} + (C_i + I_i) \frac{T}{t}$$

For the multiple items production, i.e., for $i = 1, 2, \ldots, n$ items, the total cost for the planned duration is as follows:

$$E = \frac{t}{2} \sum_{i=1}^{n} (1 - D_i / y_i P_i) D_i H_i + \frac{T}{t} \sum_{i=1}^{n} (C_i + I_i) \qquad (4.20)$$

Optimal value of t can be obtained by differentiating Eq. (4.20) with respect to t and equating to zero:

$$\frac{dE}{dt} = \frac{1}{2} \sum_{i=1}^{n} (1 - D_i / y_i P_i) D_i H_i - \frac{T}{t^2} \sum_{i=1}^{n} (C_i + I_i) = 0$$

$$\text{Or } t^2 = \frac{2T \sum_{i=1}^{n} (C_i + I_i)}{\sum_{i=1}^{n} (1 - D_i / y_i P_i) D_i H_i}$$

$$\text{Or } t = \sqrt{\frac{2T \sum_{i=1}^{n} (C_i + I_i)}{\sum_{i=1}^{n} (1 - D_i / y_i P_i) D_i H_i}} \qquad (4.21)$$

In order to illustrate, consider the two items with a total planned duration:

$$T = 36 \text{ months}$$

Table 4.13 shows the parameters for both the items, including the proportion of acceptable components in a production batch.

TABLE 4.13

Parameters for Both Items

Parameters	$i = 1$	$i = 2$
Fixed setup cost C_i	₹1570	₹1202
Apportioned innovation cost I_i	₹1500	₹1200
Total demand D_i	360	540
Inventory holding cost H_i	₹16	₹16
Total production capacity P_i	1200	2700
Proportion of acceptable components y_i	0.95	0.90

From Eq. (4.21), the production cycle time can be obtained:

$$t = \sqrt{\frac{2 \times 36 \times 5472}{(3941.0526 + 6720)}}$$

Or $t = 6.08$ months

In the context of multiple products, perfect as well as imperfect batch for purchase and production scenario has been considered along with necessary details.

5 Conclusion

Conventionally, inventory planning is discussed for purchase as well as production inventory. However, in recent times, innovation activities are given due significance in business or industrial organizations. This book especially covers innovation aspects in inventory planning and the focus is on costs. Thus, the innovation efforts in procurement and manufacture of products are highlighted so that it becomes easier to estimate the cost pertaining to the innovation efforts. Such innovation costs are incorporated in the planning for inventories where the inventories refer to the purchase and production domain in an organization. Before discussing the scope and benefits of the approach, the concluding remarks are also provided.

5.1 REMARKS

The important results are also summarized while concluding the book chapters. These are provided as follows:

(i) Innovation efforts play a significant role in the related inventory planning and the associated activities of an organization.

(ii) It is possible to link the innovation efforts with the costs, namely, the innovation cost for an organization.

(iii) An apportioned innovation cost needs to be linked to the other conventional inventory cost parameters in order to arrive at the total relevant cost.

(iv) The total relevant cost should correspond to an appropriate planning horizon concerning a specific innovative product or a group of items. After optimization of such total cost, a suitable cycle time can be obtained. This cycle time has a significant role in inventory planning with innovation.

(v) The cycle time needs to be revised with an inclusion of permissible back-orders in the purchase activity.

(vi) The production cycle time also needs to be revisited with an inclusion of permissible shortages.

(vii) An innovative product entry as well as exit should be analyzed in a multiple items situation.

5.1.1 SINGLE-ITEM PURCHASE

There is an established importance of purchase or procurement by a trading firm as well as a manufacturing concern. Some of the results are summarized as follows:

(a) Cycle time for procurement in months is given as follows:

$$t = \sqrt{\frac{2T(C+I)}{DH}}$$

where
  T = total planned duration in months
  C = fixed ordering cost
  I = apportioned innovation cost for one cycle
  D = total demand for the planned duration
  H = inventory carrying cost per unit for one month

(b) Significance of $\left(\dfrac{D}{T}\right)$ ratio can be underlined while observing its variation.

$$\% \text{ decrease in cycle time } t = 100\left[1 - \sqrt{\frac{1}{(1+L/100)}}\right]$$

where:

  L = % increase in $\left(\dfrac{D}{T}\right)$
Also,

$$\% \text{ increase in } t = 100\left[\sqrt{\frac{1}{(1-L/100)}} - 1\right]$$

where:

  L = % reduction in $\left(\dfrac{D}{T}\right)$

(c) It may be advantageous to maintain the procurement cycle time. In the context of restoring the cycle time,

$$M = \frac{L}{(1+L/100)}$$

where:

  L = % increase in $\left(\dfrac{D}{T}\right)$
  M = % reduction in H

Also,

$$M = \frac{L}{(1-L/100)}$$

where:

  L = % reduction in H

$$M = \% \text{ increase in} \left(\frac{D}{T} \right)$$

(d) Considering a reduction in $\left(\frac{D}{T} \right)$ ratio, and in the context of restoring the cycle time,

$$M = L\left[1 + \frac{C}{I}\right]$$

where:

$L = \%$ reduction in $\left(\frac{D}{T} \right)$

$M = \%$ reduction in I

Also,

In the context of implications of $\left(\frac{C}{I} \right)$,

$$M = L\left[1 + \frac{1}{(1 + N/100)}\right]$$

where:

$$I = C\left(1 + \frac{N}{100}\right)$$

Also,

$$M = L\left[1 + \frac{1}{(1 - N/100)}\right]$$

where:

$$I = C\left(1 - \frac{N}{100}\right)$$

(e) For restoring the cycle time:

$$M = L\left(1 + \frac{C}{I}\right)$$

where:
$L = \%$ reduction in H
$M = \%$ reduction in I

Also,

$$M = \frac{L}{1 + (C/I)}$$

where:

    L = % reduction in I

    M = % reduction in H

(f) Procurement cycle time with permissible backorders is given as follows:

$$t = \sqrt{\frac{2T(H+K)(C+I)}{HDK}}$$

where K = stock out cost per unit for one month
For restoring the t value,

$$M = \frac{(C+I)HL}{I\{H+K(1-L/100)\}}$$

where:

    L = % reduction in K

    M = % reduction in I

Also,

$$M = \frac{HL}{K+(H+K)(L/100)}$$

where:

    L = % increase in K

    M = % reduction in H

5.1.2 MULTI-ITEM PURCHASE

When multiple items are procured in a trading/manufacturing firm, certain results are summarized.

(a) Cycle time for procurement in months is given as follows:

$$t = \sqrt{\frac{2T\sum_{i=1}^{n}(C_i + I_i)}{\sum_{i=1}^{n}D_iH_i}}$$

where:

    $i = 1, 2, ..., n$ items

    T = total planned duration in months

    C_i = fixed ordering cost for an item i

    I_i = apportioned innovation cost for one cycle for an item i

    D_i = total demand for an item i for the planned duration (T)

    H_i = inventory carrying cost for an item i per unit for one month

(b) In the context of innovative product entry and restoring the cycle time:

$$H_j = \frac{2T \sum_{i=1}^{n}(C_i + I_i) - t^2 \sum_{\substack{i=1 \\ i \neq j}}^{n}(D_i H_i)}{D_j t^2}$$

In the context of an idea being generated for the suitable innovation cost:

$$I_j = \frac{t^2 \sum_{i=1}^{n}(D_i H_i) - 2T \left\{ \sum_{i=1}^{n}(C_i) + \sum_{\substack{i=1 \\ i \neq j}}^{n}(I_i) \right\}}{2T}$$

(c) In the context of innovative product exit and restoring the cycle time:

$$\sum_{i=1}^{n}(I_i) = \frac{t^2 \sum_{i=1}^{n}(D_i H_i) - 2T \sum_{i=1}^{n}(C_i)}{2T}$$

In order to explore the feasibility of one innovative product j,

$$I_j = \frac{t^2 \sum_{i=1}^{n}(D_i H_i) - 2T \left\{ \sum_{i=1}^{n}(C_i) + \sum_{\substack{i=1 \\ i \neq j}}^{n}(I_i) \right\}}{2T}$$

(d) With reference to a variation in the planning horizon,

$$M = 100 \left[1 - \sqrt{(1 - L/100)} \right]$$

where:
  $L = \%$ reduction in T
  $M = \%$ reduction in t

Also,

$$M = 100 \left[\sqrt{(1 + L/100)} - 1 \right]$$

where:
  $L = \%$ increase in T
  $M = \%$ increase in t

5.1.3 SINGLE-ITEM PRODUCTION

When single item is produced by a manufacturing concern, some of the results are summarized as follows:

(a) Cycle time for production in months is given as follows:

$$t = \sqrt{\frac{2T(C+I)}{(1-D/P)DH}}$$

where:

T = total planned duration in months
C = fixed setup cost per facility setup
I = apportioned innovation cost per production cycle
D = total demand for the planned duration (T)
P = total production capacity for the total duration T
H = inventory holding cost per product in ₹/month

(b) In the context of a variation in the innovation cost,

$$M = 100\left[\sqrt{\frac{C+I(1+L/100)}{(C+I)}} - 1\right]$$

where:

L = % increase in I
M = % increase in production cycle time

Also,

$$M = 100\left[1 - \sqrt{\frac{C+I(1-L/100)}{(C+I)}}\right]$$

where:

L = % reduction in I
M = % reduction in production cycle time

(c) In the context of restoring the t value,

$$M = \frac{IL}{(C+I)}$$

where:

L = % reduction in I
M = % reduction in H

Also,

$$M = \frac{L(I+C)}{I}$$

where:

  L = % reduction in H

  M = % reduction in I

(d) In the context of a variation in T and restoring the production cycle time,

$$M = \frac{100L(P-D)}{100D + L(P-D)}$$

where:

  L = % reduction in T

  M = % reduction in P

Also,

$$M = \frac{100L(P-D)}{100D - L(P-D)}$$

where:

  L = % increase in T

  M = % increase in P

(e) With reference to the increase in T and restoring the production cycle time, there can be a joint variation in the innovation cost and the production capacity, and

$$N = 100(C+I)\left[\frac{(1-D/P)(1+L/100) - \{1 - D/P(1+M/100)\}}{I(1-D/P)(1+L/100)}\right]$$

where:

  L = % increase in T

  M = % increase in P

  N = % reduction in I

(f) Production cycle time with permissible shortages is given as follows:

$$t = \sqrt{\frac{2T(H+K)(C+I)}{HDK(1-D/P)}}$$

where K = shortage cost per unit for one month.

5.1.4 MULTI-ITEM PRODUCTION

When multiple items are produced in a manufacturing firm, certain results are summarized.

(a) Cycle time for production in months is given as follows:

$$t = \sqrt{\frac{2T \sum_{i=1}^{n} (C_i + I_i)}{\sum_{i=1}^{n} (1 - D_i / P_i) D_i H_i}}$$

where:

$i = 1, 2, \ldots, n$ items
T = total planned duration in months
C_i = fixed setup cost per facility setup for an item i
I_i = apportioned innovation cost per production cycle for an item i
D_i = total demand of an item i for the planned duration (T)
P_i = total production capacity for an item i for the total duration T
H_i = inventory holding cost per item for an item i in ₹/month

(b) In the context of an inclusion of innovative item and restoring the cycle time,

$$H_j = \frac{2T \sum_{i=1}^{n} (C_i + I_i) - t^2 \sum_{\substack{i=1 \\ i \neq j}}^{n} (1 - D_i / P_i) D_i H_i}{t^2 (1 - D_j / P_j) D_j}$$

In the context of an idea being generated for the suitable innovation cost,

$$I_j = \frac{t^2 \sum_{i=1}^{n} (1 - D_i / P_i) D_i H_i - 2T \left\{ \sum_{i=1}^{n} (C_i) + \sum_{\substack{i=1 \\ i \neq j}}^{n} (I_i) \right\}}{2T}$$

(c) In the context of an exclusion of innovative item and restoring the cycle time,

$$\sum_{i=1}^{n} (I_i) = \frac{t^2 \left\{ \sum_{i=1}^{n} (1 - D_i / P_i) D_i H_i \right\} - 2T \sum_{i=1}^{n} (C_i)}{2T}$$

In order to explore the feasibility of one innovative product j,

$$I_j = \frac{t^2 \left\{ \sum_{i=1}^{n} (1 - D_i / P_i) D_i H_i \right\} - 2T \left\{ \sum_{i=1}^{n} (C_i) + \sum_{\substack{i=1 \\ i \neq j}}^{n} (I_i) \right\}}{2T}$$

5.2 SCOPE

The present discussion in the book has a wide scope in terms of application. These range from single to multiple products in purchase as well as production situations. An inclusion of quality aspects makes it more realistic. Permissible backorders in the procurement and production activities need revisiting the formulation and analysis. This section also includes a future scope suitably. The discussion refers to

(i) Within the organization
(ii) Beyond the organization

5.2.1 WITHIN THE ORGANIZATION

In a trading organization, a certain way is proposed to assess the innovation efforts. Since the main aim is to sell the innovative product to the potential customers finally, the assessment belongs to the concerned activities. These activities may include in the context of innovation efforts:

(a) Identification of potential customer
(b) Communication with such a customer
(c) Convincing the customer for the innovation component of a product
(d) Price-related discussion
(e) Resulting final sales and billing

The proposed way of looking at such aspects would lead the firm toward critical and analytical observation. This observation refers to the time taken by the resources in order to result the final sales in addition to the methodology adopted.

In case where it is a manufacturing or industrial organization, the related functions need to be carefully understood, as shown in Figure 5.1. This understanding helps a lot in a precise estimation of innovation efforts and the resulting costs among other concerned aspects of the associated and interrelated activities.

Research, design, and development usually play a significant role in starting the manufacture of an innovative product. Their efforts should be observed and utilized in a suitable estimation of the innovation cost. An input also comes for the procurement of materials in this context from the design function.

Procurement of the appropriate material in order to produce an innovative item then happens. Specific additional efforts should be accounted in the present context. The procured material is then processed on the identified facilities and the newer activity must not be missed. Else an erroneous cost estimate might result.

Similarly, the quality aspects need to be observed and the impact on the acceptance or rejection of items must be taken into consideration. The finance and accounts function has a role in the release of appropriate budget for the overall innovation activities among other tasks.

An imperfect lot can also be related to quality problems. These have scope in terms of a revised cycle time among other aspects. The results are summarized in

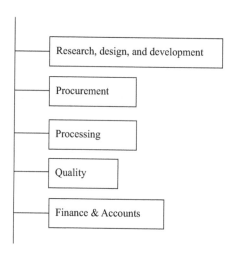

FIGURE 5.1   Related functions.

the context of procurement as follows, particularly when unacceptable components are added to the inventory level:

(a) For single-item purchase:

Procurement cycle time $t = \sqrt{\dfrac{2Ty(C+I)}{DH}}$

where y = proportion of acceptable components in a lot

(b) For multi-item purchase:

$$t = \sqrt{\dfrac{2T\displaystyle\sum_{i=1}^{n}(C_i+I_i)}{\displaystyle\sum_{i=1}^{n}(D_iH_i / y_i)}}$$

where y_i = proportion of acceptable components in a lot for an item i

The results are also summarized in the context of production as follows, particularly when unacceptable components are not added to the inventory level:

(i) For single-item production:

Manufacturing cycle time $t = \sqrt{\dfrac{2T(C+I)}{(1-D/yP)DH}}$

where y = proportion of acceptable components in a production batch

(ii) For multi-item production:

$$t = \sqrt{\dfrac{2T\sum_{i=1}^{n}(C_i + I_i)}{\sum_{i=1}^{n}(1 - D_i / y_i P_i)D_i H_i}}$$

where y_i = proportion of acceptable components in a production batch for an item i

5.2.2 BEYOND THE ORGANIZATION

Many activities are beyond the organizational boundaries such as inbound logistics and outbound logistics. Practices vary depending on the specific case and location among other factors. Transportation and storage practices of input as well as output items need to be changed in the specific case depending on the innovative item requirements. Additionally, a significant interaction may also be required with outside agencies for different functions:

(i) Design and development
(ii) Supplier development and relationship
(iii) Sales

In order to design a component supplied by another company, it might be necessary to explain/discuss the precise requirement in the context of a potentially modified component. Such factors should be reflected in the estimation of innovation cost. Future scope also lies in case studies concerning this, because the insights/implications vary according to the following:

(a) Industry
(b) Item
(c) Cost/profit scenario
(d) Investment level

The outcome in the context of estimation of costs may vary because the interaction with another company might be

(i) Negligible
(ii) Considerable
(iii) Extensive

Thus, the relevant activities and efforts often go beyond the organization under consideration.

Supplier development and relationship also depends on the current problem being faced by the organization under consideration. Quality problems may be within the organization; however, these may often extend and come under the category of "beyond the organization." In addition to the quality issues, such problems also relate to the workmanship and the facility/maintenance among others. Because of the quality problems, either the defective items might be found at the buyer location in a lot, or these may be discarded at the supplier location also. However, the cost implications still need to be taken into consideration in the event of defective items being discarded at the supplier location itself. Production/quality problems might occur owing to the facility/maintenance also, and thus need to be analyzed explicitly in the presence of such issues.

As the sales of innovative products is the final outcome in the whole chain of agencies, additional interaction efforts would be required. These efforts may include tasks such as helping the concerned agency in the identification of potential customers and training the agency members. This training might be required in making them aware of the innovation component of a product. Such awareness can relate to the technical aspects and also commercial/business/utility aspects. The discussed relevant efforts should be accounted and included in the innovation cost estimation. A more precise decision outcome is the end result if the described estimation covers all significant aspects of the situation.

After discussing the scope within the organization as well as beyond it, there are other benefits also. These benefits among others also contribute toward a better decision-making approach such as the outsourcing analysis.

5.3 BENEFITS

An overall benefit lies in a fresh approach toward linking the inventory analysis with an innovation effort and the concerned cost. In addition to the ordering and carrying cost, an innovation cost is explicitly incorporated in the purchase scenario. For the production situation also, the concerned cost with innovation is considered along with the facility setup and holding cost. Depending on the specific case, quality aspects and the permissible backorders can be included in the formulation and subsequent analysis in the procurement as well as the manufacture of items.

Some of the inventory items can be outsourced, whereas the option might be available for the company to manufacture them within the premises also. Companies at an individual level and also in the channel context would have novel opportunities to reconsider the plan for inventories, where the inventories may refer to

(i) Their own generation
(ii) Outsourced

Figure 5.2 shows a possibility of different outcomes after aligning with the proposed approach.

In case where the component is outsourced, there is a possibility of different outcomes after the implementation of the proposed approach. This might be in the context of insourcing the component if it is feasible. However, the existing practice may continue if an overall cost benefit still lies in it. On the other hand, when a

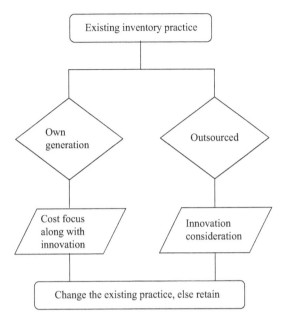

FIGURE 5.2 Possibility of different outcome.

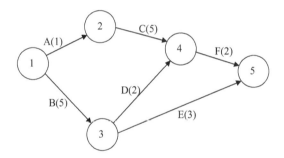

FIGURE 5.3 Activities in an innovation project.

component is insourced, i.e., the own generation, a possibility might be in the form of outsourcing it. Else the existing practice can be retained if an overall benefit still lies in it.

A potential benefit also relates to the synergy with other approaches. For example, an estimation of the innovation cost is a significant task while implementing the proposed methodology. Depending on the specific case, there is a possibility of planning for innovation in the project mode. In an innovation project, there can be multiple activities running simultaneously. And also there can be few activities which start only when a specific task is completed.

Figure 5.3 depicts various activities in an innovation project comprising six activities A, B, C, D, E, and F. Expected duration for completion of such tasks is given in weeks. This duration is provided within the bracket along with the activities. It is convenient to know the total innovation cost with the use of direct and indirect costs depending on the specific case. As per the situation, apportioned innovation cost can be estimated for the proposed application in inventory planning.

After incorporating the innovation efforts and the concerned cost in the inventory planning for a certain horizon, a procurement or production cycle time is obtained. An additional benefit also concerns the firms' relationships in the context of outsourcing. A focused firm can be associated with either trading or manufacture. Thus, a focused function can be either procurement or production depending on the case. For example, if purchase/procurement is such a function as shown in Figure 5.4, production cycle time at another firm might be synchronized directly or indirectly.

In case where a focused function is a production, the supply of input item from the other firm might be synchronized as shown in Figure 5.5.

A developed understanding of such a possible synchronization leads to superior firms' relationship/coordination as represented by Figure 5.6.

The inventory planning with innovation efforts and cost focus has been described in this book. Both procurement and production inventory are analyzed for single as well as multiple products. For the benefit of readers, it is also finally concluded with few important remarks, scope, and benefits.

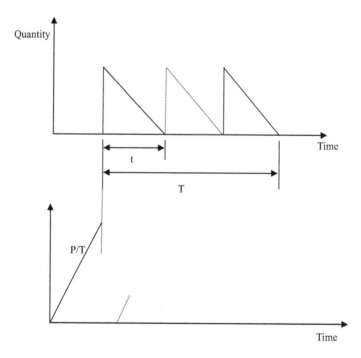

FIGURE 5.4 Purchase as focused function.

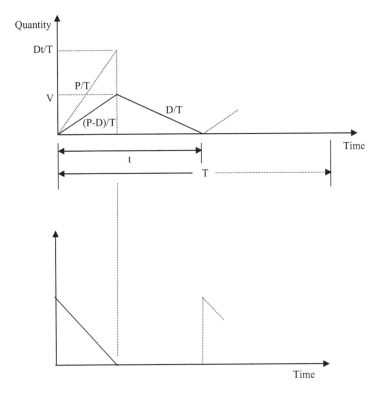

FIGURE 5.5   Production as focused function.

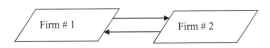

FIGURE 5.6   Firms' coordination/relationship.

General Reading I

PROJECT MANAGEMENT

Installation of plant and machinery among others may refer to projects. A project consists of several activities. Some activities may run in parallel, whereas some activities may only start when certain specific operations are over. Innovation efforts may also be planned like a project in certain cases. Utilization of resources becomes an important issue so that the project is completed on time with minimum cost. Bar chart (Figure I.1) is a simple technique for project planning in which various activities are shown with respect to time.

Activities A, B, C, and D are shown in Figure I.1. It can be determined when an activity will start and when it will end. But it is difficult to determine an interdependence among activities, particularly when these are relatively large in number.

Network methods are more useful techniques of project management as it is convenient to show interdependence among activities by drawing a network diagram. A network diagram is shown in Figure I.2.

Four activities A, B, C, and D are shown on arrows and four events 1, 2, 3, and 4 are represented by nodes. Events and activities are basic elements of the network. Resources including time are consumed in activity or operation or job. Activities are usually shown on arrows. No resources including time are consumed for the event. In Figure I.2, event 1 represents the simultaneous start of activities A and B. Event 2 represents completion of activity A as well as the start of activity C. Event 3 shows the completion of activity B and start of activity D. Event 4 is the completion of activities C and D. As the event 4 is an end event, this shows the completion of project also. For example, a network diagram can be drawn for a small project with activities A, B, C, D, E, F, and G. Activity-predecessor relationship is shown in Table I.1. A completion of E, F, and G activities can be treated as the completion of this project.

Network diagram can be constructed as shown in Figure I.3. Event 1 is the starting event. As activity A is having no immediate predecessor, it is the first activity. Event 2 marks the completion of activity A. As A is immediate predecessor to the three activities B, C, and D, these activities emanate from the event 2 and their completion is marked by the events 3, 4, and 5, respectively. Activities E, F, and G follow activities B, C, and D, respectively. Activities E, F, and G converge to an end event 6 because their completion shows the end of the project.

If an additional condition is:

Activity G cannot start unless C is completed.

Then the network diagram is drawn as shown in Figure I.4. Completion of activity C is the event 4 and the start of activity G is an event 5. Events 4 and 5 are joined by

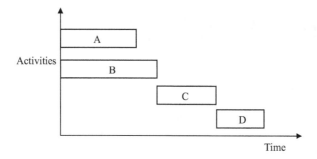

FIGURE I.1   Bar chart.

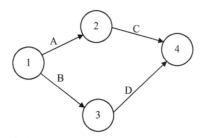

FIGURE I.2   Network diagram.

TABLE I.1
Activity-Predecessor Relationship

S. No.	Activity	Immediate predecessor
1	A	–
2	B	A
3	C	A
4	D	A
5	E	B
6	F	C
7	G	D

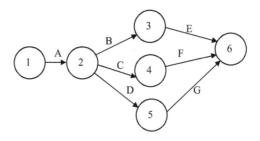

FIGURE I.3   Constructed network diagram.

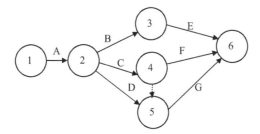

FIGURE I.4   Network diagram with dummy activity.

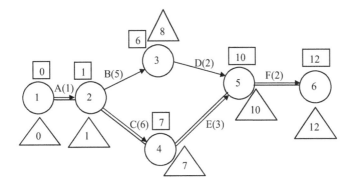

FIGURE I.5   Activity duration (days in bracket for activities).

the dotted line. 4–5 is dummy activity which satisfies the condition that activity G cannot start unless C is completed or the event 4 is over. Dummy activity does not consume resources including time. It is just to show a dependence among activities.

CRITICAL PATH METHOD (CPM)

CPM is activity-oriented technique of project management. It is suitable for the projects where the duration of the activities can be estimated precisely and there is certainty in the activity duration. Therefore, it is appropriate for deterministic situations and there is only one time estimate for each activity. Consider the network diagram as shown in Figure I.5 with activity duration.

Earliest finish time (*EFT*) corresponding to the starting event 1 is zero represented by rectangle.

EFT for any activity = largest *EFT* of predecessor activities + activity duration

EFT for activity A = 0 + 1 = 1, as there are no predecessors to A, and the activity duration of A is 1 day.

Since predecessor of B is A only, *EFT* for activity B = *EFT* of A + duration of B = 1 + 5 = 6.

Similarly, *EFT* for C = 1 + 6 = 7.

EFT for D = 6 + 2 = 8.

EFT for E = 7 + 3 = 10.

There are two predecessors to F, i.e., D and E.

*EFT*s for D and E are 8 and 10 days, respectively. Largest *EFT* of predecessor activities is 10, i.e., the largest among 8 and 10 days. *EFT* for F = largest *EFT* of predecessor activities + duration of F = 10 + 2 = 12 days.

For convenience, *EFT* is given in Figure I.5. Corresponding to node 5, out of 8 and 10, a maximum value 10 is written in rectangle. To calculate *EFT* for each activity, a forward movement is made with reference to Figure I.5.

Latest finish time (*LFT*) corresponding to the last event = *EFT* corresponding to that event. In the triangle, 12 is written to specify *LFT* corresponding to the event 6.

LFT for any activity = *LFT* of successor activity – duration of successor activity.

If there are more than one successor to any activity, then the smallest value of *LFT* thus calculated would be chosen as *LFT* for that activity.

As F is the last activity, *LFT* for F = *EFT* = 12 days.

To calculate *LFT* for all the activities, a backward movement would be made corresponding to Figure I.5.

LFT for D = *LFT* of successor, i.e., F – duration of F = 12 – 2 = 10.

As the successor to E is also F,

LFT for E = 12 – 2 = 10.

In the triangle, 10 is mentioned to represent *LFT* for convenience in Figure I.5.

LFT for B = *LFT* of D – duration of D = 10 – 2 = 8.

Similarly, *LFT* for C = *LFT* of E – duration of E = 10 – 3 = 7.

There are the two successors to A, namely, B and C.

Considering B, *LFT* for A = 8 – 5 = 3.

Considering C, *LFT* for A = 7 – 6 = 1.

Out of 3 and 1, the smallest value, i.e., 1 will be chosen as the *LFT* for A which is written in Figure I.5 corresponding to node 2.

For event 1 which is the starting event, obviously 0 can be mentioned. Now,

Total float = *LFT* – *EFT*

Activities for which total float is zero, i.e., *LFT* = *EFT*, are called critical activities, and the path having these critical activities is known as critical path.

TABLE I.2
EFT and *LFT* for Activities

Activity	EFT	LFT	TF
A	1	1	0
B	6	8	2
C	7	7	0
D	8	10	2
E	10	10	0
F	12	12	0

EFT and *LFT* as well as total float (*TF*) for each activity are given in Table I.2. Critical activities are A, C, E, and F, therefore the critical path is A–C–E–F, shown as double arrows in Figure I.5. Total project duration = 12 days.

Now,

Earliest start time (*EST*) for any activity = *EFT* for that activity – activity duration.

Latest start time (*LST*) = *LFT* – activity duration.

For example, duration of activity D = 2,

EST for D = *EFT* for D – duration of D = 8 – 2 = 6 days

LST for D = *LFT* for D – duration of D = 10 – 2 = 8 days

Similarly, *EST* and *LST* are obtained for all the activities, as given in Table I.3.

Total float (*TF*) can also be defined as (*LST* – *EST*). In addition, independent float and free float are defined as follows:

(a) *Independent float:* It is the duration by which an activity may be delayed with no effect on the floats of predecessor activities.
(b) *Free float:* It is the duration by which an activity may be delayed with no effect on the floats of successor activities.

In the previous example, activities are shown on the arrows. Consider another diagram as shown in Figure I.6 with activities on nodes and also their duration. Activities 1, 2, 3, 4, 5, and 6 have been shown on the nodes.

TABLE I.3

***EST* and *LST* for Activities**

Activity	Activity Duration	EST	LST
A	1	0	0
B	5	1	3
C	6	1	1
D	2	6	8
E	3	7	7
F	2	10	10

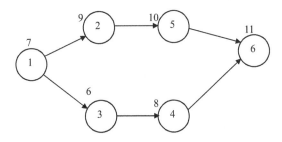

FIGURE I.6 Network diagram with activities on nodes.

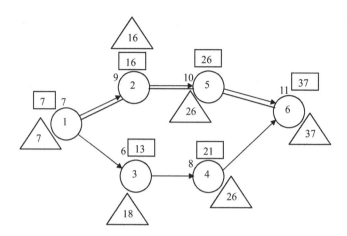

FIGURE I.7 Critical path for the network diagram.

Calculations for *EFT* of all the activities are made while moving forward as discussed before. *LFT* is obtained while moving backward. The nodes on which *EFT* and *LFT* are equal (i.e., *TF* = 0) are joined by double lines to mark the critical path as represented by Figure I.7.

Activities 1, 2, 5, and 6 are critical activities and the critical path is with an overall completion time of the project as 37.

Critical path is the longest path in terms of its duration. If activities on the critical path or the critical activities are delayed, the entire project gets delayed. All the resources should be utilized efficiently in order to complete the project on time. There is a slack of 5 on noncritical activities as *TF* is 5 corresponding to them. Even if such activities are delayed because of some unavoidable reasons, project completion time will not suffer. If need arises, resources may also be transferred from noncritical activities to the critical activities. Due to some reasons, if an overall project duration is to be shortened, then the duration of critical activities need to be shortened with an additional direct cost.

PROGRAM EVALUATION AND REVIEW TECHNIQUE (PERT)

PERT is an event-oriented technique. It is suitable for projects where duration of activities cannot be estimated precisely, and there is uncertainty in activity duration. It is appropriate for a probabilistic situation, and there are three time estimates for each activity:

(i) *Pessimistic time:* This is based on the pessimistic approach, such as the resources are not available as and when required. Obviously, this is the maximum time estimate.
(ii) *Optimistic time:* This is based on the optimistic approach, such as all the resources are available as and when required. Obviously, this is the least time estimate for completion of an activity.

(iii) *The most likely time estimate:* This is at the intermediate level between the pessimistic and optimistic time approach. Its frequency of occurrence is the highest.

Beta distribution may represent the frequency diagram for these time estimates. Now,

$$\text{Variance} = \left[\frac{(T_p - T_o)}{6} \right]^2$$

$$\text{Expected time, } T_e = \frac{(T_o + 4T_m + T_p)}{6}$$

where:
T_o = optimistic time estimate
T_m = most likely time estimate
T_p = pessimistic time estimate

From the three time estimates, an expected time (T_e) is calculated. In order to obtain the critical path, the rest of the procedure is similar to CPM. Earliest expected time and latest allowable time are obtained for each event, as the *EFT* and *LFT* were obtained in CPM.

Slack = latest allowable time − earliest expected time
Slack is zero corresponding to the events available on the critical path.

Consider the three time estimates for each activity in days, as shown in Figure I.8. For activity, 1-2:

$$T_e = \frac{(T_o + 4T_m + T_p)}{6}$$

$$= \frac{3 + (4 \times 6) + 15}{6}$$

$$= 7$$

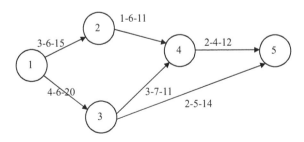

FIGURE I.8   Three time estimates for activities.

TABLE I.4

Expected Time and Variance for Activities

Activity	Expected Time	Variance
1-2	7	4.00
1-3	8	7.11
2-4	6	2.78
3-4	7	1.78
3-5	6	4.00
4-5	5	2.78

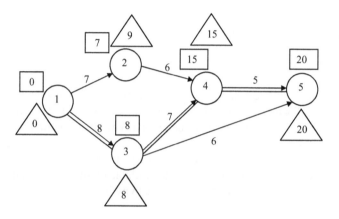

FIGURE I.9   Expected time for activities.

Variance for activity, 1-2:

$$\left[\frac{(T_p - T_o)}{6}\right]^2$$

$$= \left[\frac{(15-3)}{6}\right]^2$$

$$= 4$$

Similarly, the expected time and variance for each activity are obtained as provided in Table I.4.

Expected time for each activity is given in the network diagram, as shown in Figure I.9, and the critical path is also marked. The critical path, i.e., the longest path is 1-3-4-5 with an expected total project duration as 20 days.

Critical activities and their variance are provided in Table I.5.

Thus, the variance of critical path = 11.67.

TABLE I.5
Variance for Critical Activities

Critical Activities	Variance
1-3	7.11
3-4	1.78
4-5	2.78

And standard deviation, $\sigma = \sqrt{11.67} = 3.41$

In order to understand the concept of probability for completion of project, consider the normal distribution with the two extremes, i.e., the expected value $\pm 3\sigma$, i.e., $Z = \pm 3$, where Z is known as normal deviate. Probability associated with Z as –3 and +3 is 0.13 and 99.87%, respectively. For other intermediate values of Z between –3 to +3, available probabilities may be used.

Expected critical path duration or total project duration = 20, which is the central value. Probability of completion of project in this duration, i.e., 20 = 50% corresponding to $Z = 0$.

For example, if probability of completion of project in 22 days is to be obtained, then

$Z = \dfrac{22 - 20}{3.41}$, and corresponding to this value of Z, probability may be known. Thus,

$$Z = \frac{S_t - E_t}{\sigma}$$

where

S_t = scheduled time against which probability is to be known
E_t = expected time of completion of project corresponding to the critical path

Probability of completion of project in the expected critical path duration is 50% because

$Z = \dfrac{20 - 20}{3.41} = 0$, and for $Z = 0$, probability is 50%. With respect to normal distribution, as we move on the positive side of Z, probability is more than 50%. On the negative side of Z, the probability is less than 50%.

CRASHING OF ACTIVITIES

Sometimes a project needs to be completed earlier than the prescribed time. This affects the project costs. Total cost of the project may consist of two components:

(i) *Direct cost:* These are the costs which can be directly associated with the activities. If activity duration is reduced, then the resources are required at a higher cost and therefore the direct cost increases.

(ii) *Indirect cost:* This is related to an overhead cost among others which depends on an overall project duration, i.e., the critical path duration. If the critical path duration is reduced due to reduction in the activity time, then the indirect cost decreases.

Crashing of activities refers to a reduction in the duration of activities on the critical path so that an overall project duration is reduced. This is also called crashing of the network.

Cost slope which is the ratio of increase in cost due to crashing of an activity and a decrease in time of activity is an important parameter and should be calculated for the activities. Critical activities are identified and the critical activity with the minimum cost slope is crashed. After each crashing, the critical path may change and needs to be determined again. The process of crashing is repeated till optimum duration (i.e., with minimum total cost) of the project is achieved. If a network is to be crashed to a certain limit, then the effect on the cost can also be known with this method.

Consider the network diagram as shown in Figure I.10.

Direct cost along with the duration in days for each activity is given in Table I.6.

Assume indirect cost as ₹50 per day. Now the crashing method may be applied to find the network with minimum cost. Now:

$$\text{Cost slope} = \frac{\text{Increase in cost}}{\text{Decrease in duration}}$$

For example, in case of the activity C, the cost slope $= \dfrac{530-500}{6-3}$

$$= \frac{30}{3}$$

$$= ₹10 \text{ per day}$$

Similarly, the cost slopes of all the activities are obtained and provided in Table I.7.

Total cost needs to be obtained without crashing after knowing the critical path as shown in Figure I.11.

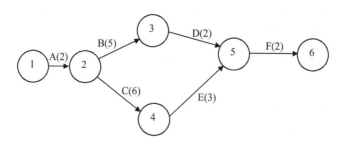

FIGURE I.10 Network diagram with activity duration.

TABLE I.6
Direct cost along with duration for activities

	Normal		Crash	
Activity	Duration	Cost (₹)	Duration	Cost (₹)
A	2	180	1	280
B	5	390	4	450
C	6	500	3	530
D	2	200	1	400
E	3	300	2	380
F	2	150	1	300

TABLE I.7
Cost Slope for Activities

Activity	Cost Slope (₹/Day)
A	100
B	60
C	10
D	200
E	80
F	150

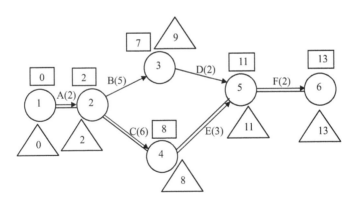

FIGURE I.11 Critical path without crashing.

Total cost = direct cost + indirect cost
Direct costs = 180 + 390 + 500 + 200 + 300 + 150 = ₹1720
As the overall project duration is 13 days,
Indirect costs = 13 × 50 = ₹650
And, the total cost = 1720 + 650 = ₹2370

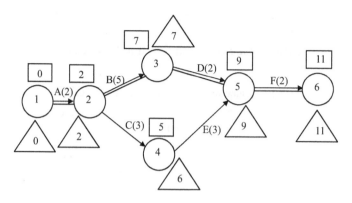

FIGURE I.12   Critical path with crashing.

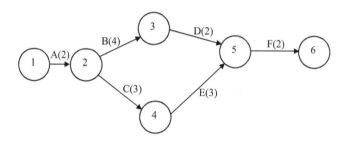

FIGURE I.13   Network diagram with further crashing.

Now the critical activities are A, C, E, and F with the cost slope as 100, 10, 80, and 150, respectively. Activity C has the minimum cost slope, i.e., 10, and it is to be crashed first.

Crash duration for C is 3, and with this duration, the critical path is obtained as shown in Figure I.12.

Increase in direct cost due to crashing = 530 – 500 = 30

Now, direct cost = 1720 + 30 = ₹1750

Indirect cost = 11 × 50 = ₹550

Total cost = 1750 + 550 = ₹2300

Now, the critical activities are A, B, D, and F with the cost slope as 100, 60, 200, and 150, respectively. Activity B is to be crashed now as it has the minimum cost slope. The obtained network diagram with further crashing is shown in Figure I.13.

It may be observed that all the activities have become critical along with the overall project duration as 10 days. Direct cost is increased by ₹60 and indirect cost is reduced by ₹50.

Direct cost = ₹1810

Indirect cost = 10 × 50 = ₹500

Total cost = 1810 + 500 = ₹2310

Direct cost slope is ₹60 per day, whereas indirect cost is ₹50 per day. As an increase in direct cost is more than the reduction in indirect cost, the total cost has increased in comparison with the previous stage.

Therefore, the minimum cost network is as shown in the previous stage (Figure I.12) with the total cost as ₹2300 and project duration as 11 days. However, following the same procedure, an effect on the total cost can be determined by crashing the remaining activities.

EXERCISES

1. What is the limitation of a bar chart that is overcome in network methods?
2. Differentiate between activity and event.
3. Explain the dummy activity.
4. Construct the network and determine the critical path considering the following details:

Activity	Predecessor	Duration
A	–	3
B	A	1
C	A	3
D	A	2
E	B	6
F	C	1
G	C, D	5
H	E, F, G	2

Also obtain the following for each activity:
   (a) *EFT*
   (b) *LFT*
   (c) *EST*
   (d) *LST*
   (e) Total float
5. Write the basic difference between CPM and PERT.
6. Discuss the three time estimates regarding PERT. How the expected time and variance are to be calculated?
7. Determine the critical path for the network diagram as given below:

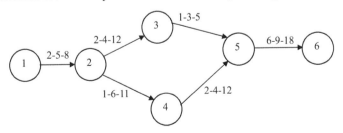

Also obtain the variance for each activity. And calculate the variance of the critical path. Comment on the probability of completing the project in expected critical path duration.

8. What do you understand by the crashing of activities? Explain the procedure by taking a suitable example.

General Reading II

FINANCIAL MANAGEMENT

In an industrial or business organization where a huge amount of money is involved and stakes are very high, the significance of financial management is much more. Financial management deals with proper planning of overall expenditure, how money is to be spent, whether a proposal is to be accepted, and priority decision in addition to numerous other functions. Apart from financial management, the following are the important functional divisions in an industrial organization:

(a) *Manufacturing:* Its aim is to convert raw materials into finished goods of a well- defined quality by efficient use of resources. For simplicity, one can include quality control, maintenance, personnel, stores, and purchase under the major heading "manufacturing."
(b) *Marketing:* Its purpose is sales and distribution of finished goods. In addition to this, the marketing division also provides input data for demand forecasting which will be used later to generate the production plan.
(c) *Accounts:* In addition to preparation of balance sheets and other routine functions such as record-keeping, it is also involved in product cost estimation and process cost accounting.

As one can understand, the integration of various activities always brings fruitful results. Financial management alone cannot function satisfactorily. It depends on other divisions: "manufacturing," "marketing," and "accounts." This interdependence is shown in Figure II.1.

This will be clearer if we come across the following problems which are often encountered:

(i) "Manufacturing" desires that the raw material should be in abundance so that production processes never stop for want of material and the resources are not idle. Similarly, "Marketing" requires a huge inventory of finished goods so that these are supplied to the customers as and when desired. At the same time, "Finance" views inventory as a burden because the capital is blocked for a considerable period. This is a reason why a trade-off would be required.
(ii) Manufacturing is concerned with achievement of production targets satisfactorily. For this purpose as well as for high quality requirements, newer and sophisticated machines are periodically required. The question is whether it is justified with the present production level.

FIGURE II.1   Interdependence of financial management.

(iii) As time passes, there is wear on the machine parts. Maintenance may not be an easy and cost-effective task. There is a proposal for replacement of the machine. The question is whether it is to be accepted.

(iv) An industrial organization is in the process of diversification or is involved in setting up of a new plant. There are alternatives for similar equipment. Which alternative should be selected?

(v) An ISO 9000 organization is dealing with a particular product. Due to stringent quality requirements and total quality management (TQM) approach, emphasis is on the production of parts with "zero defects." This firm is not using an old facility for a few months because it generates more number of defects. Suddenly the demand of the parts increases. To take advantage of this situation for profit maximization, should the company not use the old facility since an increase in the production capacity is the need of the hour?

(vi) If a proposal is accepted, then there is a need to invest money. Various sources of finance are as follows:
 (a) Banks and financial institutions
 (b) Issuance of bonds
 (c) Preference/equity shares
 Cost of different sources of raising capital is to be estimated and a judicious mix is to be adopted.

(vii) By offering the credit to customers for some period, the sales revenue may increase. But a financial analysis needs to be carried out to determine the net effects of credit policy.

(viii) Accounting method of the cost has a bearing on financial analysis. Sales price of a product is determined by precise estimation of product cost which in turn depends on the process cost accounting among other factors. Further it is of utmost interest for the management to know the break-even quantity which is the sales volume corresponding to the no-profit, no-loss situation.

From the above discussion, it is clear that several proposals/situations queue up before finance professionals for decision/acceptance. This process is summarized in Figure II.2.

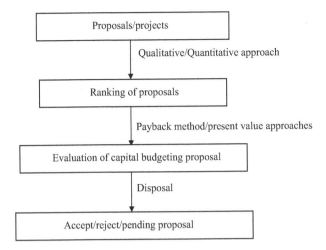

FIGURE II.2   Financial analysis of proposals.

Usually the projects are accepted on the basis of profitability by their evaluation. Annual cash flows need to be projected for evaluation of capital budgeting proposals. Determination of these cash flows largely depends on demand forecasting. Forecasting of demands of the product or services is an essential step in financial analysis. A thorough study of economic situations also helps in the demand analysis. Macroeconomics and microeconomics have an impact on the financial analysis because the environment in which organizations work affects the decision-making. Certain amount of risk is always involved in investment decisions because forecasts involve error. Statistical measures are useful in evaluating the risk such as standard deviation.

Yet another important aspect of financial management is the time value of money; for instance, if a person comes across the two proposals: (i) receipt of ₹1000 today, and (ii) receipt of ₹1000 after one year. Obviously, even a layman can tell that the first proposal is preferable because the money obtained today is more valuable than the similar amount to be received later. Therefore, the amount of money alone cannot be analyzed unless time factor is associated with that. Many of the financial analytical studies are primarily centered at the present value of future expenditures and receipts.

II.1   TIME VALUE OF MONEY

As discussed before, money in absolute terms cannot be compared. Rather, it is to be analyzed in relation to the time factor. From our day-to-day experience also, if we deposit any amount P in the bank, then an amount F which is more than P is expected after n number of years. This is shown in Figure II.3.

Conventionally, if money goes from our pocket, i.e., an expenditure or investment is made, it is shown on the negative side on the vertical axis. Similarly, any receipt of money or revenue is a kind of gain and is shown on the positive side in Figure II.3.

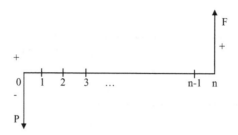

FIGURE II.3 Conventional presentation for present investment and receipt in future.

Assuming an interest rate i per year, the amount P at time 0, i.e., at present, becomes $P + Pi$, i.e., $P(1 + i)$ after one year.

At the end of two years, it will be $P(1 + i)^2$. Similarly, at the end of n number of years:

$$F = P(1+i)^n$$

$$\text{Or } P = \frac{F}{(1+i)^n} \tag{II.1}$$

where $\dfrac{1}{(1+i)^n}$ is known as the present value factor. Reciprocal of this, i.e., $(1+i)^n$ is the compound amount factor.

Example 1

Work out the present value of future receipt of ₹10,000 after 10 years, considering an interest rate as 10%. Also find out the compound amount factor.
From Eq. II.1,

$$P = \frac{F}{(1+i)^n}$$

Here F = ₹10,000
n = 10 years
i = 10%, i.e., 0.1
Therefore,

$$P = \frac{10,000}{(1.1)^{10}}$$

$$= ₹3855.43$$

Also,
Compound amount factor $= (1 + i)^n$
$$= (1.1)^{10}$$
$$= 2.5937$$

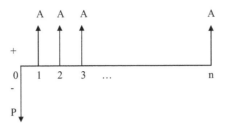

FIGURE II.4 Cash flow diagram.

ANNUITY

There are many practical situations in which an equal amount of money is deposited periodically or received periodically; for example, life insurance policy, in which we deposit equal amount of money yearly/half-yearly/quarterly for a definite period. Similarly, bank gives loan to a person and the person has to repay the loan yearly in equal amounts for a certain number of years. This kind of situation is shown in Figure II.4. In this cash flow diagram, A is called an annuity, which is a series of periodic receipts or payments of equal amount.

Now, from Eq. II.1,

Present value (PV) corresponding to A at the end of year $1 = \dfrac{A}{(1+i)}$
where i is as usual, i.e., the interest rate in %.

Similarly, PV corresponding to A at the end of year $2 = \dfrac{A}{(1+i)^2}$

And PV for A at the end of year $n = \dfrac{A}{(1+i)^n}$

The present value P of the annuity is the sum of above values, i.e.,

$$P = \frac{A}{(1+i)} + \frac{A}{(1+i)^2} + \frac{A}{(1+i)^3} + \cdots + \frac{A}{(1+i)^n}$$

$$\text{Or} \quad P = \frac{A}{(1+i)}\left[1 + \frac{1}{(1+i)} + \frac{1}{(1+i)^2} + \cdots + \frac{A}{(1+i)^{n-1}}\right]$$

The value in parentheses corresponds to well-known geometric progression, $a, ar, \ldots, ar^{n1}$, and sum of this series is

$$a\frac{(r^n - 1)}{(r-1)}$$

By application of this,

$$P = A\left[\frac{(1+i)^n - 1}{i(1+i)^n}\right]$$

$$\text{Or } P = A\left[\frac{1-(1+i)^{-n}}{i}\right] \tag{II.2}$$

Value in bracket is called annuity present value factor. Reciprocal of this, i.e., $\left[\dfrac{i}{1-(1+i)^{-n}}\right]$, is known as loan repayment factor or capital recovery factor (CRF).

Annuity may also continue forever, and in such a case, $n \to \infty$ and from Eq. II.2:

$$P = A\left[\frac{1-(1+i)^{-\infty}}{i}\right]$$

$$\text{Or } P = A\left[\frac{1-0}{i}\right]$$

$$\text{Or } P = \frac{A}{i} \tag{II.3}$$

Such type of annuity is known as perpetual annuity.

Example 2

An institution is in the process of developing a fund from which medals may be given to the toppers every year. It is also estimated that an average expenditure of ₹12,000 will be incurred annually. Incidentally, an old student of the institution who is a successful professional now wishes to deposit the amount at once for this purpose. How much should he donate to the institution considering an interest rate of 12% so that the medals may be given for a very long time?

Eq. II.3 is applicable as this concerns the perpetual annuity. Thus,

$$P = \frac{12000}{0.12}$$

$$= ₹100,000$$

Therefore, the old student of the institution should be advised to donate ₹100,000 for this purpose.

Example 3

A person decides to deposit ₹10,000 at the end of every year in his bank account for 15 years. How much should he receive at the end of 15 years considering an annual interest rate of 10%?

This situation is depicted in Figure II.5.

Value at the end of year 1 = 10,000
Value at the end of year 2 = 10,000 + 10,000 (1 + 0.1)
Value at the end of year 3 = 10,000 + 10,000 (1.1) + 10,000 (1.1)^2

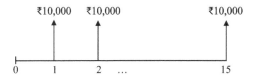

FIGURE II.5 Representation of annual deposit.

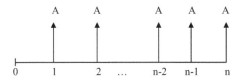

FIGURE II.6 General approach with uniform annual deposit.

Value at the end of year 15 = $10,000 + 10,000 (1.1) + 10,000 (1.1)^2 + \cdots + 10,000 (1.1)^{14}$

$$= 10,000\, [1 + 1.1 + (1.1)^2 + \cdots + (1.1)^{14}]$$

$$= 10,000 \left[\frac{(1.1)^{15} - 1}{1.1 - 1} \right]$$

$$= ₹317,724.82$$

Following a general approach, this can be shown in Figure II.6.
Future value at the end of n number of years,

$$F = A + A(1+i) + A(1+i)^2 + \cdots + A(1+i)^{n-1}$$

$$\text{Or } F = A \left[\frac{(1+i)^n - 1}{i} \right]$$

where $\left[\dfrac{(1+i)^n - 1}{i} \right]$ is called as annuity compound amount factor.

Reciprocal of this, i.e., $\left[\dfrac{i}{(1+i)^n - 1} \right]$, is known as sinking fund factor.

Familiarity with various factors is useful in financial analysis and planning.

Example 4

Find out (a) annuity compound amount factor, and (b) sinking fund factor for the data of Example 3.

(a) Annuity compound amount factor = $\left[\dfrac{(1+i)^n - 1}{i} \right]$

$$= \left[\frac{(1.1)^{15} - 1}{0.1} \right]$$

$$= 31.7725$$

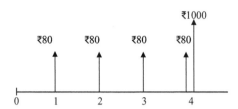

FIGURE II.7   Representation for the gain from investment.

(b) Sinking fund factor $= \left[\dfrac{i}{(1+i)^n - 1} \right]$

$= \left[\dfrac{0.1}{(1.1)^{15} - 1} \right]$

$= 0.03147$

Example 5

Consider an investment which will give ₹80 in interest per year and ₹1000 at maturity and that is at the end of four years from now. What should be the present value of this investment?

Assume interest rate as 8%.

Representation for the receipts is shown in Figure II.7.

Present value of the investment = present value of the stream of ₹80 from 1 to 4 years + present value of ₹1000 to be obtained at the end of four years.

With the use of Eqs. II.1 and II.2,

$$\text{Present value of the investment} = A\left[\frac{1 - (1+i)^{-n}}{i} \right] + \frac{F}{(1+i)^n}$$

where

A = 80
i = 0.08
n = 4
F = 1000

Now,

$$PV \text{ of the investment} = 80\left[\frac{1 - (1.08)^{-4}}{0.08} \right] + \frac{1000}{(1.08)^4}$$

$$= 264.97 + 735.03$$

$$= ₹1000$$

In this example, the present value of the investment works out to be the same as maturity value because the coupon interest rate is equal to the yield to maturity.

II.2   CAPITAL STRUCTURE

Capital structure may be related to the suitable proportion of securities issued by a firm. Securities include shares or stocks and bonds among a variety of types. These are issued for raising the capital and are used as the sources of finance. An industrial or business organization may issue bonds for general public to raise the capital for their projects. The company pays the interest to bond holder and at the end of stated number of years, returns the original amount of money also. As these are a kind of debt taken by the company, the company is liable to pay interest to the bond holders whether there is profit or not.

Broadly speaking, shares may be of two types: equity shares and preference shares. Since shareholders are considered to be the owners of the company, their original money is not refunded by the company particularly for equity shares. Equity shares are also called common stocks. Their risks are high and they are the last to have a claim over profits. Preference shares are also called preferred stocks. As the name suggests, they will get preference over common stocks, i.e., the dividends will be paid first to the preference shareholders and then to the equity shareholders.

Business organizations are engaged in production and projects among other activities. For these activities, they require finance. For raising the finance, the companies issue securities such as shares and bonds. All the sources of finance have their advantages and disadvantages. For instance, in case of debt, i.e., bond/debenture, the interest payments are tax deductible.

Suppose that a company needs ₹10,000 for its project. (*Note:* To keep the illustration simple, a very small figure is considered. In practice, it is a large amount.) Assume that the firm raises entire amount in the form of debt by issuing bonds and promises to pay the bond holders interest at the rate of 10%. Consider the expected annual earnings from the project as ₹5000 and tax rate as 30%. The calculations for projected income are shown in Table II.1.

Since the company has raised ₹10,000 in the form of debt, it is liable to pay interest @10%. This is why the interest ₹1000, i.e., 10% of ₹10,000, is deducted from the earning to get the taxable income. One can easily understand that the interest payments are tax deductible. This is an advantage of debt. On the other hand, the firm has to pay the interest even if no or less earning is there due to some reasons.

TABLE II.1
Projected Income with
Bonds as Source of Finance

Annual earning	₹5000
Interest to bond holders	₹1000
Taxable income	₹4000
Tax @30%	₹1200
Projected income	₹2800

Now consider the preference shares as the only source of finance in the present example. Fixed dividends are to be paid to the shareholders, say @15%. The calculations are shown in Table II.2.

Since only preference shares are issued, no interest payment is there and the company could not take advantage of a reduced taxable income. Dividend is 15% of 10,000, i.e., ₹1500, and the projected income is ₹2000, which is less than that shown in Table II.1.

Table II.3 shows the projected income with equity shares as the only source of finance. Assume that the equity shareholders desire 17.5% rate of return (*ROR*).

Suppose that 1000 equity shares have been issued out of which 500 are held by the original promoters. Now,

$$\text{Earnings per share (EPS)} = \frac{\text{Earning to common stock holders}}{\text{Total number of equity shares}}$$

$$= \frac{3500}{1000}$$

$$= 3.5$$

TABLE II.2
Projected Income with Preference Shares as Source of Finance

Earning	₹5000
Interest	Nil
Taxable income	₹5000
Tax @30%	₹1500
Earning after tax	₹3500
Dividends to preference shareholders	₹1500
Projected income	₹2000

TABLE II.3
Projected Income with Equity Shares as Source of Finance

Earning	₹5000
Interest	Nil
Taxable income	₹5000
Tax @30%	₹1500
Earning after tax	₹3500
Dividends to preference shareholders	Nil
Projected income to common stock holders	₹3500

In this example, it was deliberately stated that the required *ROR* for equity shares is 17.5%. Because after distributing ₹1750 (which is also evident from multiplication of *EPS*, i.e., 3.5 and 500 shares purchased by others), remaining ₹1750 would be available to the company as income. In reality, a lot of uncertainty is there in equity shares. If earnings are less, *EPS* will be less and shareholders will get less than the required *ROR*. On the other hand, in case of high earnings, shareholders will get much more than the required *ROR* since *EPS* is high. However, investor expects minimum required *ROR* and it depends on the risk involved. The company raising the finance may view it as cost of capital if capital is gained from the equity shares.

Cost of capital may be viewed as the minimum required rate of return. Industrial or business organizations raise finance from different sources and the costs of different sources of raising capital are as follows:

(a) *Cost of debt:* In the example, it is 10%, i.e., the rate of interest paid to the bond holders.
(b) *Cost of preference shares:* In the example, it is 15%, i.e., the rate at which the fixed dividend is to be paid to preference shareholders.
(c) *Cost of equity shares:* This is the reflection of minimum required *ROR* by common stock holders. In the example, it is considered to be 17.5%.

Individual sources of raising capital have been considered before. Each source has advantages and disadvantages. In practice, the capital structure consists of suitable proportions of debt, preference shares, and equity shares.

For example, the capital structure of an enterprise consists of the following:

(i) Debt of ₹200,000 at 10% interest
(ii) Preference shares of ₹100,000 at 15% dividend
(iii) 90,000 equity shares have been issued @ ₹10

Assume that the current level of *ROR* on equity shares is 17.5% and tax applicable to the enterprise is @30%. In order to find out the weighted average cost of capital or weighted average required return, total value of the capital would be needed.

Total value of capital = 200,000 + 100,000 + (90,000 × 10)
$$= ₹1,200,000$$

Now,

Proportion of debt in capital structure $= \dfrac{2}{12}$

Proportion of preferred stocks in capital structure $= \dfrac{1}{12}$

Proportion of equity stocks in capital structure $= \dfrac{9}{12}$

Since the interest payments are tax deductible, effective cost of debt considering 30% tax rate would be

$$10(1 - 0.30) = 7\%$$

Thus, the weighted average cost of capital can be obtained as follows:

$$\left[\frac{2}{12}\right] \times 7 + \left[\frac{1}{12}\right] \times 15 + \left[\frac{9}{12}\right] \times 17.5$$

$$= 15.54\%$$

II.3   CAPITAL BUDGETING

Budgeting is concerned with knowing the finances available and how to spend it. Similarly, capital budgeting is associated with the following:

(a) What kind of capital assets are to be procured?
(b) How much expenditure is allowed for procuring these capital assets?

Various proposals are evaluated by the financial analysts. Payback period is one method among various available approaches. Suppose we invest ₹10,000 in procuring an equipment and an expected annual earnings due to that is ₹2000. It can be easily said that the costs involved will be recovered in 5 years. In this process, we have divided capital investment by an expected annual earning.

Another approach is net present value (*NPV*) method. A forecast is made for future cash flows which are $A_1, A_2, ..., A_n$ as shown in Figure II.8, and the proposed investment is P for any project.

Now, the net present value can be given as

$$NPV = \frac{A_1}{(1+r)} + \frac{A_2}{(1+r)^2} + \frac{A_3}{(1+r)^3} + \cdots + \frac{A_n}{(1+r)^n} - P \qquad (II.4)$$

where r = cost of capital or the required return
In case of the annuity, and with the use of Eq. II.2,

$$NPV = A\left[\frac{1-(1+r)^{-n}}{i}\right] - P \qquad (II.5)$$

If the *NPV* is positive, then the benefits are more than the investment and the proposal is accepted.

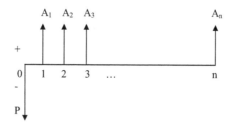

FIGURE II.8   Proposed investment with forecasted cash flow.

Example 6

An entrepreneur is considering the procurement of metallizing equipment at the initial investment of ₹100,000. Its life is 12 years and the projected cash flows per year are approximately ₹30,000 for the life of the equipment. If the cost of capital is 10%, state whether the entrepreneur should decide in favor of procurement on the basis of *NPV* method.

From Eq. II.5,

$$NPV = A\left[\frac{1-(1+r)^{-n}}{i}\right] - P$$

Here,

A = ₹30,000
r = 0.1
n = 12
P = ₹100,000

Therefore,

$$NPV = 30,000\left[\frac{1-(1.1)^{-12}}{0.1}\right] - 100,000$$

$$= 204,410.75 - 100,000$$
$$= ₹104,410.75$$

As the *NPV* is positive, the entrepreneur must decide in favor of the procurement of equipment.

Broadly speaking, the cash flow is annual revenue minus the yearly expenditure. But the net cash flow is net of taxes, i.e., taxes are deducted from the cash flow. In order to find out the taxable income, depreciation is reduced from, revenue – expenditure. In other words, the depreciation is tax deductible.

Depreciation is the reduction in value of an equipment as time passes. For example, consider the present value of an equipment as ₹14,000 and it is expected that at the end of 12 years useful life, its salvage value will be ₹2000. Then an overall depreciation for the life of equipment is (14,000 – 2000), i.e., ₹12,000 and the yearly depreciation may be considered as

$$\frac{12,000}{12}$$

$$= ₹1000$$

However, in certain cases, the firm may depreciate the equipment completely over some number of years which are less than the life of equipment.

Example 7

Assume that the depreciation per year for the plant and equipment of the company is ₹100,000. Yearly revenue and expenses are ₹1 million and ₹0.4 million,

TABLE II.4
Calculation for Net Cash Flow

Revenue (A)	₹100,0000
Expenses (B)	₹400,000
(A – B)	₹600,000
Less depreciation	₹100,000
Taxable income	₹500,000
Tax @35%	₹175,000
Net cash flow (A – B – tax)	₹425,000 (600,000 – 175,000)

TABLE II.5
NPV and Initial Investment of Projects

Project	NPV (₹)	Initial Investment
X	30,000	35,000
Y	95,000	50,000
Z	130,000	110,000

respectively. Tax rate is considered to be 35%. Table II.4 shows the calculations for net cash flow.

Usually a company has many projects which await approval. The company will obviously like to undertake the projects which yield high profit per unit investment. Since *NPV* is a measure of profitability in a way, *NPV* per unit of initial investment is called as profitability index. For example, a firm has the three proposed projects whose *NPV* and initial investment are shown in Table II.5.

From the available information, *NPV* per unit investment can be found as follows:

X: 0.86
Y: 1.90
Z: 1.18

For the purpose of knowing the order of preference, these proposals should be arranged such that the relevant index is in the decreasing order as given below:

Y: 1.90
Z: 1.18
X: 0.86

Assume that the firm has a constraint on the capital budget, i.e., it can invest only ₹60,000. In such a case, the firm should select the project Y because it is on the top priority in the present context.

Fixed assets include plant and equipment. Whenever we visit any industrial organization, the following questions come to our mind:

(a) Why this type of industry is situated at this place only?
(b) Why various departments and facilities are located in a particular order?
(c) How do they predict the customer demands and plan their production/service accordingly?
(d) What kind of techniques are adopted in order to improve the management and operational efficiencies with an objective of maximizing the profit or minimizing the costs?
(e) How do they finance various projects and other routine activities as survival of the organization becomes difficult without effective financial management?

The objective of any entrepreneur is to locate the industrial or business organization at such a place where it is convenient to run the operations and the total costs are minimized. The ideal location for any industrial organization is depicted in Figure II.9.

This is the situation in which raw materials as well as market for finished goods are near to the industry. Easy availability of labor and infrastructure, including basic facilities such as power, water, and transportation, exist. But in the real environment, ideal conditions do not always exist. The effort should be to avail maximum benefits. Numerous factors affecting plant localization are categorized as follows: (a) significant factors, (b) considerable factors, and (c) additional factors.

(a) *Significant factors*
(i) Proximity to raw materials and market

If the source of raw materials is in nearby area of industrial location, it is advantageous from the point of view of transportation cost. Similarly, if the location is closer to a marketplace from where the finished products are available to consumer, the cost of transportation can be kept to a minimum. In case either the raw material or the finished product (or both) is having a shelf life constraint, i.e. these can be spoiled in a specified time period (e.g., in food processing industry), proximity of raw materials/market becomes critical.

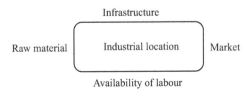

FIGURE II.9   Ideal location of an industrial organization.

(ii) Infrastructure

Efficient transportation system along with the buildup roads and smooth availability of power and water is necessary for industrial or economic activities.

(iii) Availability of labor

As long as the manpower is cheaper than automation and if it does not affect the quality of product, the availability of labor will be a major issue in some of the industrial activities.

  (b) *Considerable factors*
  (i) Availability of supplementary industries in proximity

An industry requires many components some of which are purchased from outside sources. If the ancillary units manufacturing these components are in proximity of the main industry, then it is always preferred. This factor is becoming significant day by day because of the supply chain management and just-in-time inventory systems.

(ii) Storage facilities

The criterion gets utmost attention in some cases such as the agricultural industry.

(iii) Government/taxation policy

Government may prefer to develop a backward area by encouraging industrialists to establish their setups at that place. Rebate in the taxes may be offered for a specified period.

(iv) Financial facilities

Financial institutions as well as banks are preferred in the proximity due to an ease in the management of current assets/long-term loans.

  (c) *Additional factors*

(i) Strategic factor

All the industries should not be located at one place. Rather, these must be scattered throughout the nation. Due to a natural tragedy or war, one or few regions may be destroyed and if the industries are situated at several places in the country, then the whole economy will not be ruined.

(ii) Size/price of the plot and its topography

Size and shape of the available plot is to be taken into consideration for the layout design. Depending on the situation, price of the land also becomes important. Topography, i.e., ups and downs of the region, is of concern from civil engineering point of view because of the leveling effort and cost involved.

In the context of site selection, there are basically two types of sites: (a) urban area, and (b) rural area.

(a) Urban area

Usually urban region has good infrastructure and an easy availability of professionals and skilled labor. But there is less scope for expansion of the industry. Price of the land is high and pollution is a major concern.

(b) Rural area

The problem of pollution is less. Land is cheaper and more scope for expansion of the industry exists. Industrial relations are good. But poor infrastructure and availability of professionals are limited as most of the professionals may not like to live in the rural area.

After discussing the advantages and disadvantages of urban as well as rural areas, suburban area seems to be a compromise between urban and rural regions. Quite satisfactory infrastructure is available in suburban area and professionals also do not hesitate in living there.

Different factors are discussed some of which are difficult to quantify. However, an attempt should be made to compare various alternative sites by evaluating the total costs based on input items, the resources to convert them into finished products, selling expenditure, and taxes.

The criteria are also classified as shown in Figure II.10.

Objective factors are those which can be quantified in the money value. Subjective factors are qualitative type of factors which cannot be evaluated in money terms. Critical factor is the criterion in the absence of which a particular site can never be considered.

To meet the customer demands, different types of production system may be adopted. Figure II.11 shows a broad classification of production system.

FIGURE II.10   Criteria affecting the selection of a location.

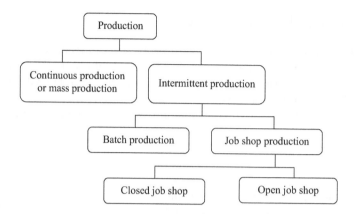

FIGURE II.11   Classification of production system.

In continuous production or mass production, the demand may be more than the production rate. The production rate is more than the demand rate in intermittent production. Job shops are of two types: closed job shop and open job shop. Example of open job shop is common welding shop or machine shop where anyone can approach for certain job to be done. Closed job shop usually caters to the needs of a particular company.

For example, a large water pump manufacturing company may have its own job shop in the same premises in addition to various departments such as casting, machining, etc. Forecasting of demands is easier for closed job shop as compared to open job shop. In job shop production, the variety of products is very high, but quantity of production for each variety is very less.

On the other hand, in mass production, quantity of production is high and a few variety of products, sometime even a single standard product, is manufactured. Therefore, considering characteristic features, such as variety and quantity of products, mass production and job shop production are at two extremes. Batch production is at the intermediate level.

Plant layout refers to the arrangement of facilities, equipment, and departments inside the industrial/business premises. The concept is useful for manufacturing as well as service type of organizations. The kinds of layout are as follows:

(a) Line or product layout
(b) Functional or process layout
(c) Fixed position layout
(d) Cellular manufacturing layout
(e) Hybrid layout

These are explained as follows:

(a) *Line or product layout*

Product layout is suitable for mass production where the volume of production is high and a variety of product is less, sometimes even a single standard product. The machines and testing facilities are arranged in the sequence of processes to be carried out and the tests at different stages. The layout is shown in Figure II.12.

The product requires first operation on machine M1, second operation on machine M2, and so on up to machine Mn. Ti/Tf represent the testing facilities at various stages. The characteristics of the product layout are as follows:

(1) Special purpose machines are installed.
(2) Semiskilled or less skilled people can run the machine, but highly skilled workers/professionals are required in order to upkeep these machines.
(3) An initial investment is high, but the operating cost and subsequently the production cost is low.
(4) If one machine stops functioning and as it is line layout, the entire production operation may come to a halt.
(5) Conveyor system for material handling can be adopted for the smooth flow of material.
(6) Work-in-progress inventory is less because the output of one machine immediately goes to the next machine if it is available to process.
(7) All the machines are specialized for a particular design of the product. If a change in design is required, then it is difficult for the production system to respond to it. Therefore, flexibility in the system is not available.

For example, the three machines exist in a product layout, as shown in Figure II.13.

Production capacity of M1 and M3 are 50 units per hour and the capacity of M2 is 25 units per hour, then M3 will be idle for half of the time. The solution may be to install two number of machine M2, as shown in Figure II.14.

This is called a perfect balance of production line if it matches the production requirement. But in the real environment, when the number of machines is more and each has different capacity of production, it is difficult to balance the line completely. Hence, line balancing is a problem to be solved in this type of layout.

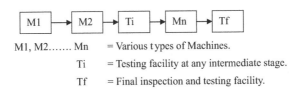

M1, M2 .... Mn   = Various types of Machines.
            Ti   = Testing facility at any intermediate stage.
            Tf   = Final inspection and testing facility.

FIGURE II.12   Line layout.

FIGURE II.13   Three machines in the product layout.

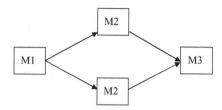

FIGURE II.14 Modified arrangement.

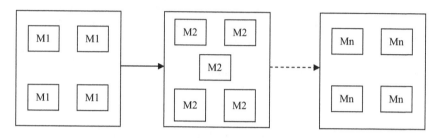

FIGURE II.15 Functional layout.

(b) *Functional or process layout*

Process or functional layout is shown in Figure II.15 in which the similar machines are grouped together at one location. However, it is not necessary for each product type to move through all types of machines. Depending on the requirements, any product may skip one or more departments.

Characteristics of this type of layout are as follows:

(1) It is suitable for an intermittent production, i.e., the job shop or batch production.
(2) General-purpose machines are installed.
(3) Highly skilled people are required to set up and run the machines.
(4) An initial investment is less, but the cost of production is high.
(5) If there is a breakdown of one or two machines, the entire production process will never be stopped.
(6) It is difficult to adopt a conveyor system for material handling.
(7) Work-in-process inventory is more.
(8) If there is change in the design of product, this type of layout is able to respond to it in a short time as the machines can be set up for a variety of products. Therefore, the system is flexible.
(9) While designing the layout, minimizing the total material handling cost is one of the main objectives.

(c) *Fixed position layout*

In a product/process layout, the material was taken to various facilities/machines. But in the fixed position layout, all the resources, including machines, are brought to materials. This is appropriate in the situation when the material and subassembly are too heavy to flow through the stationary facilities for processing. For example, in case of shipbuilding, heavy/thick steel plates are used. It is convenient to keep the plates and components together and these are joined by welding and other processes. The welding machines and other equipment as well as the workers go to the site of material and a large ship is fabricated.

(d) *Cellular manufacturing layout*

This is suitable for a group technology program. An existing job shop may select some of the parts manufactured by it which are suitable to be made in the cells. A cell acts as a product layout for such parts. A part family is a group of such parts which are having similar operational requirements, i.e. they are processed on similar machines with identical setups. Further, the demand of such parts should be relatively high so that they can be manufactured in moderate batch sizes.

(e) *Hybrid layout*

It is a combination of product and process layout.

II.4   CURRENT ASSETS

Assets are mainly of two types: fixed assets and current assets. Fixed assets or fixed capital include plant and equipment. Capital budgeting is concerned with procurement decisions of fixed assets. Such assets are generally used throughout their life span. On the other hand, current assets or working capital are mostly used in a relatively short time period such as inventories. Inventories include raw materials, purchased components, work-in-process, and finished goods.

It is difficult to pay equal attention to all the items in an industrial organization; therefore, the items are selected on the basis of certain analysis and these are classified into various categories. ABC analysis is based on the following:

(i)   Less number of items account for more value of annual usage
(ii) More number of items account for less value of annual usage

Annual usage value (or value of annual use) is the multiplication of annual consumption in units and the price per unit item.

A few items of category A refer to high annual usage value, whereas category B items account for medium value of annual use, and large number (category C items) accounts for low annual usage value.

(a) Approximately 10% of the total number of items (A items) account for approximately 70% of the total annual inventory value.

(b) About 20% of the total number of items (B items) relate to about 20% of the total inventory value.

(c) Remaining approximately 70% of the total number of items (C items) refer to about 10% of the total value of annual usage.

Classification in terms of percent is indicative only. Depending on the data, certain variations may take place in the actual categorization of the items.

Example 8

An organization wants to analyze 15 items. Annual consumption in units as well as price in ₹ per unit is given in Table II.6.

Now the annual usage value of each item is obtained by multiplying price per unit and annual consumption, as provided in Table II.7.

Now annual usage value is arranged in decreasing order, i.e., starting from the highest value, along with the corresponding item, as shown in Table II.8. Cumulative value of annual use is obtained such as 40,500 + 40,100 = 80,600, 80,600 + 10,000 = 90,600, ...

Total inventory value can be observed as ₹115,000.

Cumulative value of annual use is also obtained in terms of percent of total inventory value such as (40,500/115,000) × 100 = 35.22%, (80,600/115,000) × 100 = 70.09, ...

Out of total 15 items (i.e., from A to O), the two items G and N may be categorized as A items because these constitute for approximately 70% of the total inventory value.

TABLE II.6
Details for Various Items

Item	Price per Unit (₹)	Yearly Consumption (units)
A	4	160
B	10	110
C	5	1030
D	4	65
E	5	130
F	20	45
G	10	4050
H	20	90
I	5	2000
J	50	14
K	3	800
L	2	300
M	3	3000
N	20	2005
O	30	40

TABLE II.7
Annual Usage Value of Items

Item	Annual Usage Value
A	640
B	1,100
C	5,150
D	260
E	650
F	900
G	40,500
H	1,800
I	10,000
J	700
K	2,400
L	600
M	9,000
N	40,100
O	1,200

TABLE II.8
Calculation for ABC Classification

Category	Item	Annual Usage Value	Cumulative Annual Usage Value	Cumulative Value as % of Total Inventory Value
A	G	40,500	40,500	35.22%
	N	40,100	80,600	70.09%
B	I	10,000	90,600	78.78%
	M	9,000	99,600	86.61%
	C	5,150	104,750	91.09%
C	K	2,400	107,150	93.17%
	H	1,800	108,950	94.74%
	O	1,200	110,150	95.78%
	B	1,100	111,250	96.74%
	F	900	112,150	97.52%
	J	700	112,850	98.13%
	E	650	113,500	98.70%
	A	640	114,140	99.25%
	L	600	114,740	99.77%
	D	260	115,000	100%

$(2/15) \times 100 = 13.33\%$ of the total items (namely, G and N) are classified as A category items, which account for 70% of the total inventory value.

Three items, namely, I, M, and C, i.e., $(3/15) \times 100 = 20\%$ of the total number of items, are categorized as B items which account for $91.09 - 70.09 =$ approximately 21% of the total inventory value.

Remaining 10 items are C items which are large in number, i.e., $(10/15) \times 100 = 66.67\%$ of the total items. These C items account for $100 - 91.09 =$ approximately 9% of the total inventory value.

An item requires utmost attention, i.e., it should be reviewed frequently for their stock status and decision regarding procurement. C items which are large in number account for a small proportion of the total inventory value, and these may be purchased in bulk. They may require the least attention from the managerial point of view. B items may be considered at an intermediate level in ABC analysis/ classification.

Purchase of the item/material and their issue for various purposes is a continuous process. Methods to determine the value of closing stock/pricing the issue of material are also relevant. First-in-first-out (FIFO) method is explained with the following example:

(i) On September 9, 150 units of an item are procured @ ₹11 per unit.
(ii) On September 12, 200 units of that item are procured @ ₹13 per unit.

Now, on September 15, the item is to be issued for consumption and the quantity to be issued is 250.

On September 9, value of the stock = $150 \times 11 = ₹1650$
On September 12, the value of stock = $1650 + (200 \times 13) = ₹4250$
On September 15, 250 units of the item are to be issued. As the method is FIFO, out of 250 units:

First 150 units are issued @ ₹11 per unit = 1650
Remaining 100 units are issued @ ₹13 per unit = 1300
Value of the item issued = $1650 + 1300 = ₹2950$
Value of the stock on September 15 = $4250 - 2950 = ₹1300$

In case where last-in-first-out (LIFO) method is applied for the mentioned example:

Out of 250 units which are to be issued on September 15,
First 200 units are issued @ ₹13 per unit = 2600
Remaining 50 units are issued @ ₹11 per unit = 550
Value of the item issued = $2600 + 550 = ₹3150$
Value of the stock on September 15 = $4250 - 3150 = ₹1100$

Thus, value of the closing stock depends on the particular method to be applied. In addition to inventories, a brief overview is provided for some of the financial management aspects.

EXERCISES

1. What do you understand by the time value of money?
2. An investment gives ₹90 as interest per year and ₹1000 at maturity at the end of five years from now. Assuming an annual interest rate of 9%, evaluate the present value of this investment.
3. What do you understand by the capital structure?
4. Explain the cost of capital and the weighted average cost of capital.
5. Discuss the capital budgeting. Also explain the payback period and *NPV* method.
6. The purchase of mechanical press is being planned by a firm which costs ₹100,000. Assume that the scrap value of the press after the five years useful life is zero. The expected cash flows for five years are as follows: ₹40,000, 40,000, 40,000, 30,000, and 20,000.
This situation is shown below:

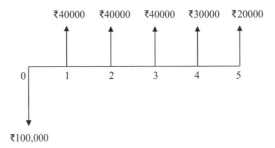

Find out:
(a) Payback period
(b) *NPV*, if the cost of capital is 12%
7. Consider the data shown in the following table for a company:

Revenue (A)	₹100,0000
Expenses (B)	₹400,000
(A – B)	₹600,000
Less depreciation	₹100,000
Taxable income	₹500,000
Tax @35%	₹175,000
Net cash flow (A – B – tax)	₹425,000

A proposal of purchasing new metallizing equipment is pending for the financial analysis whose cost is ₹100,000. The proposed equipment will improve the operations in terms of cost reduction, i.e., the expenses would be reduced by ₹40,000 per year. Assume that the annual revenue is unaffected. The equipment is to be depreciated over 8 years. Life of the equipment is 12 years.

Yearly depreciation for the proposed equipment $= \dfrac{100,000}{8}$

$$= ₹12,500$$

It is of relevance to consider the net effect on the cash flow due to the purchase of the proposed equipment. The annual depreciation of ₹12,500 will be applicable for the first 8 years. As the life of the equipment is 12 years, for the remaining 4 years, depreciation will not be considered. The following table shows the change in cash flow for the first 8 years:

Change in revenue (A)	0
Decrease in expenses (B)	−40,000
(A − B)	₹40,000
Less depreciation	₹12,500
Change in taxable income	₹27,500
Tax @35%	₹9625
Increased cash flow (A − B − tax)	₹30,375

₹30,375 is the increased cash flow for the first 8 years due to the purchase of the equipment. For the remaining 4 years, a similar type of calculation can be made except that the depreciation would be zero for the proposed addition of the equipment. Increase in the cash flow for 9–12 years is shown as follows:

Change in revenue (A)	0
Decrease in expenses (B)	−40,000
(A − B)	₹40,000
Less depreciation	0
Change in taxable income	₹40,000
Tax @35%	₹14,000
Increased cash flow (A − B − tax)	₹26,000

Generate the increased cash flow diagram. If the cost of capital is 10%, compute the *NPV* for the proposal and make a decision whether the proposal is to be accepted.

8. Why the proximity to raw materials and market is a significant factor affecting the plant location?

9. What are the advantages of having ancillary units in the proximity of the main company?

10. The industries must be scattered throughout the nation from the strategic point of view. Comment.

11. Explain the following statement "Suburban area seems to be a compromise between the urban and rural regions."
12. Discuss the objective, subjective, and critical factors affecting the selection of an industrial location.
13. State whether the line layout is suitable for shipbuilding. If not, what is the reason?
14. Differentiate between the product and process layout.
15. Under what circumstances, the cellular manufacturing layout would be suitable.
16. Assembly line balancing is an important area of the analysis of product layout. Comment.
17. What do you understand by the current assets?
18. Describe the ABC analysis in detail. Implement the ABC analysis for the following data:

Item	Price per Unit (₹)	Yearly Consumption (units)
A	3	200
B	11	100
C	5.5	1000
D	3.5	70
E	6	120
F	21	40
G	9	4100
H	19	100
I	4	2200
J	52	13
K	2	900
L	3	200
M	5	1600
N	19	2010

General Reading III

INSPECTION, QUALITY, AND PRODUCTION

Quality of any item is related to fitness for its use. It is a kind of degree of perfection. Inspection is a procedure to measure the quality of an item which may be described by certain standards and specifications. Specifications are checked by inspecting the item and it is decided whether the item is accepted or rejected.

III.1 INSPECTION

Inspection department should be organized in such a manner that the skilled staff is always available to discard the below-standard items. They need to be trained for using a variety of measuring instruments and in the upkeep of the sophisticated equipment. Different activities of the inspection division are as follows:

(a) It is important to inspect the raw materials thoroughly. If the raw material defects are detected at a later stage, entire cost of manufacturing may go waste.

(b) At every stage of manufacture, setting of the machine tools should be carefully supervised. The first produced component needs to be inspected for defects and for different specifications. Once the first component is correct in all aspects, then further production should be allowed. Work-in-progress inspection is useful because problems are detected at that level only and the defects are not passed to subsequent stages.

(c) The customers use the finished product marketed by the company. Therefore, the finished products are checked in such a way that the customers would be satisfied while using the product. In this way, the customer complaints may be minimized. However, the customer complaints cannot be eliminated completely and after receiving those, the inspection division should be able to handle them efficiently.

Inspection on the shop floor is represented by Figure III.1.

If the inspection is performed on the shop floor itself, then an inspector has to make a lot of movement. The inspector will go to the shop floor at every facility and then check the components. Sophisticated equipment may not be taken on the shop floor. But the decision-making is fast at the same time.

Centralized inspection is represented by Figure III.2.

All the items/components are taken to a specified place in the factory for the inspection purpose. Inspection takes place in suitable atmospheric conditions and modern equipment can be used. But the decision-making is slow as items/components may queue up before the centralized inspection facility.

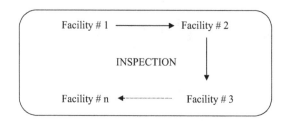

FIGURE III.1   Inspection on the shop floor.

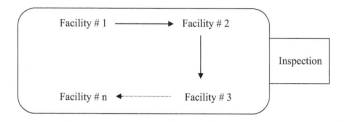

FIGURE III.2   Centralized inspection.

A combination of these two ways of inspecting may be adopted to get the advantage of both the systems.

In case of 100% inspection, all the items are checked. But in sampling inspection, out of a lot, a sample of few pieces is selected and the sample is checked for defects. If in the opinion of an inspector the sample is good enough, then the entire lot is accepted. While selecting sample, care is to be taken that the sample is a representative of the whole lot.

Suppose that in a lot of 1000 items, 100 items are picked up as sample, then the sample size is 100. Depending on the inspection of sample of 100 items, it is decided which way this lot of 1000 items goes. This is called acceptance sampling. For example, if the number of defective pieces is less than or equal to 5 after inspecting 100 pieces, the whole lot of 1000 pieces is accepted. In other words, acceptance number = 5. Since only one sample is picked up, this is called single sampling plan.

To generalize, if sample size = S, and acceptance number = A_n, then the lot is accepted if the number of defective pieces found in sample size S is less than or equal to A_n. The lot is rejected if the number of defective pieces is greater than A_n. In case of rejection, the lot may be subjected to 100% inspection and all defective pieces found are either reworked or replaced by good pieces.

In the single sampling plan, a decision is taken after inspecting one sample only. But in a double sampling plan, if need arises, a second sample is inspected. Let first sample size = S_1 and second sample size = S_2. Acceptance number corresponding to the first sample = A_{n1}. And the acceptance number for both the samples = A_{n2}.

First sample is inspected. The three situations that may arise are as follows:

(i) Number of defects $\leq A_{n1}$. In such a case, the lot is accepted.
(ii) Number of defects $> A_{n2}$. The lot is rejected in this case.

(iii) In situations (i) and (ii) above, there is no need for the second sample. But in the third case, i.e.,

Number of defects $> A_{n1}$, but the number of defects $\leq A_{n2}$. In such a case, the second sample is inspected.

If sum of the number of defects in the first and second samples $\leq A_{n2}$, the lot is acceptable. Otherwise, if the total number of defects in both the samples $> A_{n2}$, then the lot is finally rejected.

For example, $A_{n1} = 4$ and $A_{n2} = 7$.

First sample is inspected and if the defective pieces are up to 4, the lot is accepted. If defective pieces exceed 7, the lot is rejected. In case where the number of defects is 5, 6, or 7, then the second sample is inspected. If combined number of defects >7, the lot is rejected. In case where this is up to 7, the lot is accepted.

III.2 QUALITY CONTROL

For example, a component of diameter 15 mm, as shown in Figure III.3, is to be manufactured in large quantities.

As one can imagine, it is difficult to produce every component with a diameter of 15 mm exactly due to variation in the process and skill of human resources. If tolerances are provided as 0.1 mm, then efforts are made to produce items with dimensions ranging between 14.9 and 15.1 mm. However, an overall objective should be to manufacture at nominal value (i.e., 15 mm).

Kinds of variation in a process are as follows:

(a) *Unusual variation:* These are also called special cause of variation or assignable cause variation. For instance, it is identified that machine setting is not proper and due to this, variation in the process has occurred. Assignable cause variation should be eliminated completely in order to obtain a stable process.

(b) *Usual variation:* This is also called chance variation or common cause variation. Such variations are present in the system itself and responsibility for these may not lie on the employees of the company. A stable process would have only usual variation. Process improvement efforts are made when process becomes stable.

Statistical quality control (SQC) makes use of statistics in controlling the quality characteristic.

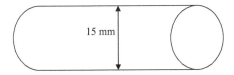

FIGURE III.3   A component for manufacture.

Variables refer to a situation when actual measurements such as 6.7 mm, 36 μm, etc. are made for each item in the sample. $\overline{X}$ and R-chart are useful for variables. For example, consider sample size = 4, i.e., in each sample, 4 items are included. Four measurements are recorded for each sample, say dimension in mm, and seven samples are collected. Measurements and calculations for $\overline{X}$ and R are given in Table III.1.

For example, in sample number 1, 4 items are measured and the readings are 15.5, 14.7, 15.2, and 14.9 mm. Out of these four observations, maximum reading = 15.5 and minimum reading = 14.7.

Range is the difference between the maximum and minimum observations, i.e.,

$$R = 15.5 - 14.7 = 0.8$$

$\overline{X}$ is the average of these four observations.

Similarly, $\overline{X}$ and R values are calculated as there are seven samples. Now,

$$\overline{R} = \frac{\sum R}{7} = 0.57$$

Similarly, $\overline{\overline{X}}$ is the average of all seven values of $\overline{X}$.

$\overline{\overline{X}} = 15.01$.

For chart, control limit = $\overline{\overline{X}} \pm A_2 \overline{R}$, i.e.,

Upper control limit = $\overline{\overline{X}} + A_2 \overline{R}$

Lower control limit = $\overline{\overline{X}} - A_2 \overline{R}$

where A_2 is a factor whose values are available corresponding to the sample size.

TABLE III.1

Measurements and Calculations

Sample No.	Measurement (mm)				Range and $\overline{X}$	
	1	2	3	4	R	$\overline{X}$
1	15.5	14.7	15.2	14.9	0.8	15.07
2	15.1	14.8	14.9	15.3	0.5	15.02
3	14.6	14.9	15.4	14.8	0.8	14.92
4	15.2	14.8	14.9	14.9	0.4	14.95
5	15.3	15.2	15.3	14.8	0.5	15.15
6	14.7	14.9	15.4	14.8	0.7	14.95
7	15.1	15.1	14.8	14.9	0.3	14.98
					$\overline{R} = 0.57$	$\overline{\overline{X}} = 15.01$

For sample size = 4, A_2 = 0.73

Upper control limit (UCL) = 15.01 + (0.73 × 0.57) = 15.42

Lower control limit (LCL) = 15.01 – (0.73 × 0.57) = 14.59

$\overline{X}$-chart is constructed as shown in Figure III.4.

For R-chart, UCL = $D_4\overline{R}$ and LCL = $D_3\overline{R}$, where D_3 and D_4 are factors whose values are available corresponding to the sample size. For sample size = 4, D_3 = 0, and D_4 = 2.28.

UCL = 2.28 × 0.57 = 1.2996

LCL = 0 × 0.57 = 0

R-chart is constructed as shown in Figure III.5.

$\overline{X}$ and R values are well within the control limits. If for one or more samples, the values fall outside the control limits, then further investigation becomes necessary. Reasons for the process going out of control occasionally should be identified and a remedy is to be suggested for implementation.

Attributes refer to a situation in which actual measurement is not taking place. The item is acceptable on the basis of certain criteria:

(i) Less than five number of scratches on the body of automobile.
(ii) Unevenness of weld bead at less than or equal to the two points on the complete weld bead.

In addition to above, finding out the number of defective pieces in a sample using "Go" and "No Go" gauges may also come under the category of attributes. Total number of defects including scratches, dents, etc. on each product may also be recorded and analyzed.

C-chart is suitable for a situation in which the number of defects are counted on each product. For example, in a steel cupboard, defects may be scratches or improper

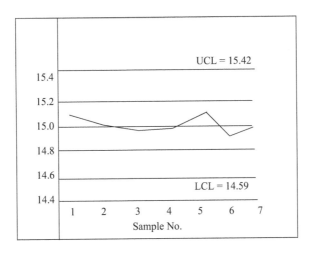

FIGURE III.4 $\overline{X}$- chart.

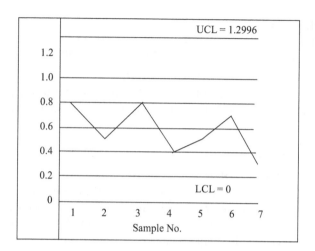

FIGURE III.5   R- chart.

TABLE III.2

Observations for Steel Cupboards

Item No.	Number of Defects, C
1	6
2	7
3	4
4	5
5	3
6	2
7	4
8	6
9	5
10	6
11	3
12	1
13	2
14	2

---

paint at one or more number of places. Fourteen steel cupboards have been considered as samples and the number of defects is counted on each cupboard. Observations are provided in Table III.2.

Total number of defects = 56

Mean number of defects, $\bar{C} = \dfrac{56}{14} = 4$

Control limit for C-chart is

$$\bar{C} \pm 3\sqrt{\bar{C}}$$

i.e., $4 \pm 3\sqrt{4} = 4 \pm 6$

UCL $= 4 + 6 = 10$

LCL $= 4 - 6 = -2$

Lower control limit is considered as zero, because negative number of defects may not exist. Now LCL $= 0$ and UCL $= 10$. C-chart is plotted as shown in Figure III.6.

P-chart is appropriate for a situation in which the samples of some items are inspected and % defectives are obtained in each sample. Item is considered to be defective if it does not match the specifications.

For example, consider sample size $= 150$.

Each sample includes 150 items. All 150 items in each sample are inspected and it is recorded that how many of them are defective items. Data collected for 12 samples and % defectives calculation is given in Table III.3.

For sample No.1, % defective, P is obtained as $\dfrac{3}{150} \times 100 = 2\%$

Similarly, P values are obtained for each sample. Average P is obtained as follows:

$$\overline{P} = \frac{\sum P}{12} = 2.44\%$$

$\overline{P}$ may also be calculated as:

(Total number of defectives)/(total number of items inspected)

That is,

$$\overline{P} = \frac{44}{(150 \times 12)} = 0.0244 = 2.44\%$$

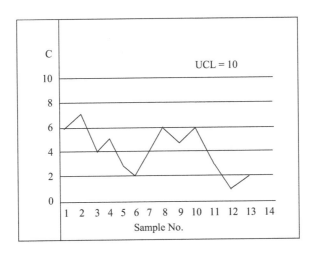

FIGURE III.6 C-chart.

TABLE III.3

% Defectives for Each Sample

Sample No.	Number of Defectives Out of 150 Items Inspected	% Defective (P)
1	3	2.00
2	4	2.67
3	2	1.33
4	5	3.33
5	3	2.00
6	4	2.67
7	1	0.67
8	6	4.00
9	7	4.67
10	3	2.00
11	4	2.67
12	2	1.33

Control limits for P-chart are

$$\overline{P} \pm 3\sigma_p$$

where $\sigma_p = \sqrt{\dfrac{\overline{P}(1-\overline{P})}{N}}$

With N = sample size = 150

$$\sigma_p = \sqrt{\frac{0.0244(1-0.0244)}{150}} = 0.0126$$

Control limits are:

$$0.0244 \pm (3 \times 0.0126)$$

UCL = 0.0622 = 6.22%

And LCL = −0.0134, i.e., LCL = 0 as negative fraction defectives may not exist.

P-chart may be constructed by plotting either fraction defectives or % defectives with respect to the sample number. Figure III.7 shows the P-chart using % defectives.

UCL and LCL determined for the process are different from the tolerance specified for the product. If process variation can be reduced, it becomes more capable to produce acceptable products.

III.3   PRODUCTION PLANNING AND CONTROL

Production planning and control has two components: (i) production planning, and (ii) production control. The planning for production-related activities is made

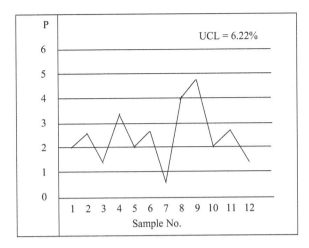

FIGURE III.7   P-chart.

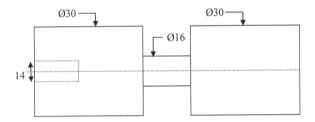

FIGURE III.8   A part to be manufactured.

at aggregate level as well as at shop floor level. For the production control, the activities during execution are supervised closely. If there is any deviation from the planned target, suitable action is taken either to eliminate the deviation or to minimize it.

Overall planning is done for producing specific goods in specific quantities with due consideration of profitability. Cost-volume-profit analysis is useful in knowing how much quantity should be produced for earning certain profit. An initial production plan is prepared and it is compared with the available capacity. When the production plan can be achieved by making use of the available capacity after a certain iterative procedure, then it is finalized for further details.

For effective utilization of the resources on the shop floor such as machines, storage space, human resources, raw materials, and tools, production planning and control (PPC) becomes necessary. Some of the aspects relevant to the PPC are discussed as follows:

Routing: This refers to the route followed by the part to be manufactured on the shop floor. In other words, it is the sequence of machines through which the part will move for different operations. For example, a part to be manufactured is shown in Figure III.8.

At the ends, the diameter of the part is 30 mm, whereas in the middle it has the diameter of 16 mm, and at one end, it requires a drilled hole of diameter 14 mm. Application of the part also suggests the grinding of the surface. Thus, the following operations are needed to be performed:

  (i) Turning
 (ii) Drilling
(iii) Grinding

Sequence of operations suggests the part to move from lathe machine through drilling machine to the grinder. The path to be moved by the product or part is decided and the routing refers to it.

Scheduling: It relates to time. It will state the time period in which a specified quantity of the product/part is to be manufactured. Depending on the practice of the company, schedule may be generated with respect to each machine/worker. The relevant worker should be aware of the target, i.e., how much quantity is to be produced, in which time period (say day or week), and on which machine. The scheduling is also useful in knowing the load on machine or worker.

The routing and scheduling are basically the planning tools. As the planned work has to be converted into physical reality, all necessary orders for release of raw materials as well as tools are issued. An objective of dispatching all necessary resources to right places is to complete the scheduled work successfully.

To control the work, effective follow-up is essential. It is to be supervised that the planned work is completed or not. If the work is in progress, whether it is being performed at the desired rate. A check is necessary for deviation in the target. Any deviation is analyzed thoroughly and remedial measure is implemented to ensure the completion of targeted production in time.

As mentioned before, the cost-volume-profit analysis is useful in knowing how much quantity should be produced for earning certain profit. Quantity to be manufactured and sold is an important decision for production planning at aggregate higher level. Planned production should be in accordance with the desired profit level. Cost-volume-profit analysis is useful for this purpose in addition to its other applications. Fixed and variable costs are considered, whereas total cost is the sum of fixed and variable costs. Fixed cost is concerned with the investment made in relevant capital assets such as plant and machinery, whereas the variable cost is related to actual manufacturing and other associated cost.

Let

Fixed cost $= f$
Variable cost per unit $= v$
Quantity to be produced $= q$

Now,

Variable cost $= v \cdot q$

Total cost $= f + vq$

If sales price per unit is s, then the sales revenue $= s \cdot q$

Profit earned, p = sales revenue − total cost

$$\text{Or } p = sq - (f + vq)$$

$$\text{Or } p + f = q(s - v)$$

$$\text{Or } q = \frac{p + f}{s - v}$$

For example, consider the relevant fixed cost of a company as ₹200,000. Selling price per unit product is ₹14 and the variable cost to manufacture this item is ₹9 per unit. If the company wants to earn a profit of ₹75,000, it is of interest to know how much quantity it should produce and sell. Now,

p = ₹75,000
f = ₹200,000
s = ₹14
v = ₹9

$$q = \frac{p + f}{s - v}$$

$$= \frac{75,000 + 200,000}{14 - 9}$$

$$= \frac{275,000}{5}$$

$$= 55,000 \text{ units}$$

If profit is zero, i.e., $p = 0$:

$$q = \frac{f}{s - v}$$

This is called as break-even quantity. At break-even point (*BEP*), there is no profit-no loss because the total cost is equal to sales revenue. With the use of given data:

$$BEP = \frac{200,000}{14 - 9}$$

$$= 40,000 \text{ units}$$

BEP in terms of ₹ = 40,000 × 14 = ₹560,000

Cost-volume-profit analysis along with its other applications helps considerably in planning for the production and related activities.

As per the demand forecast, sales revenue and total cost as well as profit may be projected for future periods and viability of the existing as well as the planned investments may be verified. Demand forecasts are also used for the production planning as well as capacity planning. Depending on the demand forecast and production economy, a production schedule is generated which will state the production quantity in respective periods. Capacity may be in terms of machine capacity, human resources, and storage space among others. If available capacity is less than the production plan requirements, then a decision is to taken whether extra capacity must be arranged or the production plan should be altered in accordance with the available capacity.

Forecasting means assessment of the future. All individuals as well as organizations are interested in the knowledge of the future, i.e., assessment of the future periods. Demand forecasting is associated with predicting the demands of product or services. There are situations when sufficient data regarding the past or the history of product demand are available. Quantitative methods are useful in analyzing such data for forecasting of the demand. In time series analysis, the data related to actual demand are plotted against time and appropriate conclusions are drawn. Various patterns of demand may be constant, increasing, and decreasing with respect to time. Accordingly, future demands may be projected extending the line further. However, there can be cyclic demand pattern. After certain time (might be many years also), the cycle repeats itself. In case of seasonal demand, this cycle completes in one year.

In the moving average method, the number of periods for which average of demands will be computed is decided by the planners. For example, if four-period average is to be considered and if at the end of the current period i, forecast f_i for period $(i + 1)$ is to be made, then

$$f_i = \frac{d_i + d_{i-1} + d_{i-2} + d_{i-3}}{4}$$

where d_i, d_{i1}, d_{i2}, and d_{i3} are actual demands in period i, $i-1$, $i-2$, and $i-3$, respectively, and f_i is the forecast made at the end of period i for the next period $(I + 1)$. At the end of period $(i + 1)$, if actual demand occurred is d_{i+1}, then

$$f_{i+1} = \frac{d_{i+1} + d_i + d_{i-1} + d_{i-2}}{4}$$

In this way, recent four periods will always be considered in this example for computing average and the most distant demand will be dropped for each successive forecast. For computing f_{i+2}, d_{i2} is not needed and will be dropped. d_{i+2} will enter the calculation which is the most recent actual demand data. In order to illustrate, assume the following:

Week	1	2	3	4	5	6
Demand (units)	5	4	6	7	7	5

Now considering four-period moving average, forecast at the end of week 6 for the next, i.e., week 7, is

$$f_6 = \frac{d_6 + d_5 + d_4 + d_3}{4}$$

$$= \frac{5 + 7 + 7 + 6}{4}$$

$$= 6.25 \approx 6$$

Forecast made in week 6 for the next week is 6 units. But at the end of week 7, actual demand is observed to be 7:

Week	1	2	3	4	5	6	7
Demand (units)	5	4	6	7	7	5	7

Now,

$$f_7 = \frac{d_7 + d_6 + d_5 + d_4}{4}$$

$$= \frac{7 + 5 + 7 + 7}{4}$$

$$= 6.5 \approx 7$$

As the period for taking average move with time, that is why the name "moving average" method of forecasting.

In order to understand weighted moving average, consider the following demand data:

Week	1	2	3	4	5	6
Demand (units)	3	5	4	2	1	3

Say, the weightage provided to the most recent data is 40%, then 30%, 20%, and 10% to consecutive older data considering 4-week computation (sum of weightage is 100%):

Week	6	5	4	3
Demand (units)	3	1	2	4
Weightage (%)	40%	30%	20%	10%

Weighted moving average = $(3 \times 0.4) + (1 \times 0.3) + (2 \times 0.2) + (4 \times 0.1)$
$$= 2.3$$

That is considered as forecast f_6 for week 7. As discussed in case of the moving average, at the end of week 7, actual demand occurred in that week will be used in computation and the oldest data, i.e., for week 3 will be left out. In moving average, equal weightage was provided to all specified period data. But in the present method, weightage is provided depending on the prevailing situation. Sometimes older data is given more weightage than the recent data.

In order to understand the exponential smoothing method, let

$$\alpha = \text{smoothing constant (value between 1 and 0)}$$

Forecast f_i for the next period $(i + 1)$ is given as

$$f_i = \alpha d_i + (1 - \alpha) f_{i-1}$$

For example, if $\alpha = 0.25$, then $f_i = 0.25 d_i + 0.75 f_{i-1}$

In other words, a forecast made for the next period is the weighted average of current actual demand and the previous forecast.

If two variables are correlated, then with the help of regression model, suitable relationship is obtained between those variables. Consider x as the advertisement expenditure and y as the sales revenue. Assuming a linear relationship:

$$y = a + bx$$

where:
Independent variable $= x$
Dependent variable $= y$

a and b are constants which need to be obtained from the past available data.

Line $y = a + bx$ represents the forecasts for various values of x. At any value x_i, y_{ia} is the actual data, whereas y_{if} is the forecast based on $y = a + bx$.

Deviation from actual $= y_{ia} - y_{if}$

Mean square deviation is a measure of error:

$$\sum_{i=1}^{n} \frac{\left(y_{ia} - y_{if}\right)^2}{n}$$

where n = number of actual data available.

To minimize the error, $\sum (y_{ia} - y_{if})^2$ should be minimized.

Say, $M = \sum (y_{ia} - y_{if})^2$

As $y_{if} = a + bx_{ia}$

$$M = \sum (y_{ia} - a - bx_{ia})^2$$

(x_{ia}, y_{ia}) are available data for independent and dependent variables and $i = 1, 2, 3, \ldots, n$.

To minimize M, differentiating it partially with respect to a and equating to zero,

$$\frac{\partial M}{\partial a} = \sum 2(y_{ia} - a - bx_{ia})(-1) = 0$$

$$\text{Or } \sum y_{ia} = na + b \sum x_{ia} \qquad \text{(III.1)}$$

Differentiating M partially with respect to b and equating to zero,

$$\frac{\partial M}{\partial b} = \sum 2(y_{ia} - a - bx_{ia})(-x_{ia}) = 0$$

$$\text{Or } \sum x_{ia} y_{ia} = a \sum x_{ia} + b \sum x_{ia}^2 \qquad \text{(III.2)}$$

Eqs. III.1 and III.2 are used to evaluate a and b, and then the linear relationship is obtained. Regression may be known as extrinsic method in which external factors related to the demand are also associated. Remaining quantitative methods, namely, time series analysis, moving average, weighted moving average, and exponential smoothing may be known as intrinsic methods, which rely completely on the history of demand of the product.

For example, a firm has the data available for advertisement expenditure and its sales revenue as shown in Table III.4.

Now, $n = 8$, and in order to use Eqs. III.1 and III.2, the related calculations are shown in Table III.5.

From Eqs. III.1 and III.2,

$$129 = 8a + 36b$$
$$\text{And } 627 = 36a + 176b$$

TABLE III.4
Available Data

S. No.	Advertisement Expenditure ₹ ($\times 10^4$)	Sales Revenue ₹ ($\times 10^6$)
1	3	14
2	4	15
3	3	11
4	5	18
5	4	12
6	6	20
7	7	26
8	4	13

TABLE III.5
Related Calculation

x_{ia}	y_{ia}	$x_{ia} * y_{ia}$	x_{ia}^2
3	14	42	9
4	15	60	16
3	11	33	9
5	18	90	25
4	12	48	16
6	20	120	36
7	26	182	49
4	13	52	16
$\sum x_{ia} = 36$	$\sum y_{ia} = 129$	$\sum x_{ia} y_{ia} = 627$	$\sum x_{ia}^2 = 176$

On solving:

$$a = 1.18$$
$$b = 3.32$$

And the relationship is obtained as

$$y_{if} = 1.18 + 3.32 x_{ia}$$

For instance, if the firm is desirous of investing ₹9 × 10⁴ for advertising its product next time, the sales revenue can be estimated as follows:

$$y_{if} = 1.18 + (3.32 \times 9)$$

$$= 31.06$$

Therefore, for advertisement expenditure of ₹9 × 10⁴, the forecast is made for sales revenue and the estimated sales revenue is ₹31.06 × 10⁶.

In addition to the discussed quantitative methods, there are certain qualitative methods also. The qualitative methods are useful when either past data are not available or insufficient data are available. However, in order to reach a judicious conclusion, a combination of quantitative and qualitative methods may be applied.

Suppose that there is a plan to launch a new product or service, then historical analogy may be found to know the growth of the demand. Similarly, if newer technology substitutes for older technology, then the rate of growth of the old technology may be treated as a basis to forecast the demand pattern of the new technology. It may safely be assumed that either in full or in part, the growth pattern may have a bearing on the historical analogies. There may be a need to search for the analogous products and their history of demand. For example, black and white TV owners will sooner or later switch to the colored TV sets. Similarly, landline telephone consumers may have a trend toward shifting to the use of mobile telephones. In case where demand forecast for mobile telephones is desired, landline telephones may be considered as an analogous product.

In the context of product life cycle, there is the relationship between demand of a product and time. In the introduction region, the product is initially launched in the market and its demand rises at a slow rate. In the growth region, demand increases at a faster rate. In the maturity zone, the product becomes mature and demand is at its peak and almost constant. As the product has its life span, finally the demand may start declining. This may be due to the following reasons:

(i) Product becomes obsolete.
(ii) Competitors have started playing an effective role and other models of the product have appeared in the market.

Before a decline in the demand of a product occurs, diversification might be needed. An analogy may be, "if the road ahead is difficult, then there is diversion." In order to avoid decline, design of the product may be modified. Utility as well as reliability of the product may be enhanced. Additional features are incorporated in the product to create the demand.

Market research is a method by which the perception of the customer about a product is known. In the form of well-designed questionnaires, information is obtained from existing as well as potential customers. After data collection, the facts are analyzed and conclusions are drawn. After knowing the liking/disliking as well as by incorporating the suggestions for improvement, demand of the product may be created or enhanced.

Sales persons represent various regions in the country and they are in close contact with the customers. Depending on the customer's requirement in future, each sales person estimates the demand of the respective region. In the sales force composite, these estimates are consolidated in order to develop the total demand forecast.

In another approach, a committee may be formed by the management to forecast the demand, and the decision of the management is given. The committee may comprise experts from various disciplines such as production, marketing, and financial management. The objective of the method is to obtain different views and then in comparatively less time, generate the forecasts.

In the Delphi technique, a panel of experts is formed. But these experts never meet in person. A coordinator seeks the response of all the experts by correspondence. Forecasts received in this manner are examined by the coordinator. If all the forecasts are in closer range, then with less effort, an appropriate decision is made. Usually forecasts may be in a wide range and these are to be brought in a narrow range by repeated queries. Consider a hypothetical example of steel-making company. A panel of seven members is formed. Some of them are from inside the organization and some are experts from outside. An experienced coordinator sends the designed format to all seven members. The forecast is desired from each along with reasons to substantiate their level of forecast. Forecasts obtained are provided in Table III.6.

After receipt of the forecasts, the coordinator compiles all the major reasons/assumptions of the members of panel. It is also observed that the member 2 has made

---

TABLE III.6
Various Forecasts

Member	Forecast ($\times 10^5$ MT)
1	2.92
2	1.90
3	3.08
4	3.05
5	2.95
6	4.00
7	3.00

TABLE III.7
Various Forecasts Including the Revised

Member	Forecast ($\times 10^5$ MT)
1	2.92
2	2.97
3	3.08
4	3.05
5	2.95
6	3.00
7	3.00

---

the forecast on an exceptionally lower side. The forecast of the member 6 is exceptionally high. Reason advanced by the member 2 is as follows: a competitor is in the process of major expansion and due to that the market share may go down. This reason along with others compiled by the coordinator is sent to the member 6 and it is asked whether in the light of new information (particularly expansion of the competitor), a revision in the forecast is possible. Now with the recent information gained, a revised forecast by the member 6 is 3×10^5 MT and it is communicated by him to the coordinator.

The reason advanced by the member 6 with his earlier forecast, i.e., 4×10^5 MT was as follows: due to government policy and change in economic scenario, it has become easier to export the steel and the demand may go high. This as well as other reasons is communicated to the member 2 whose present forecast is 1.90×10^5 MT and a query is made whether there is any possibility of upward revision particularly due to export potential. The new information is taken into account by the member 2 and the revised forecast by him is now 2.97×10^5 MT. Remaining members have also considered such reasons, but they are of opinion that decrease due to one reason and increase due to other reason are similar and therefore their forecast is same as before. Now the forecasts are as shown in Table III.7.

The coordinator now finds that the forecasts are in closer range of $(2.92 - 3.08) \times 10^5$ MT. Thus, the final forecast may be 3×10^5 MT. In real practice, the process will be repeated several times in order to create a consensus.

However, an overall understanding of the economic as well as business environment, directly or indirectly, helps in demand forecasting. Such forecasts are used as an input for the production planning of a company.

EXERCISES

1. What is inspection? Discuss different activities related to inspection.
2. Differentiate between:
   (a) Centralized inspection and inspection on the shop floor
   (b) Sampling inspection and 100% inspection
3. Describe:
   (i) Single sampling plan
   (ii) Double sampling plan
4. What are different kinds of variation in process?
5. Discuss the benefits of stable process.
6. (a) Explain how to construct $\bar{X}$ and R-charts.
   (b) For which type of situations, P-chart and C-chart are useful.
7. Compute the control limits and construct the $\bar{X}$ and R-charts for the following data related to 10 samples with sample size = 5:

Sample No.	Measurement (mm)				
	1	2	3	4	5
1	7.7	7.3	7.6	7.4	7.9
2	8.5	8.0	7.9	7.6	8.2
3	8.1	7.8	8.3	7.7	8.0
4	7.9	8.0	7.8	8.3	8.2
5	8.3	8.1	7.7	7.8	8.1
6	8.2	8.2	7.9	8.0	7.8
7	7.8	8.1	7.7	7.5	7.8
8	8.4	8.3	7.6	7.5	7.5
9	7.7	8.1	8.3	7.9	7.8
10	8.0	8.3	8.4	7.6	7.9

Consider the following factors: $A_2 = 0.58, D_4 = 2.11,$ and $D_3 = 0$.

8. Number of defects have been recorded for the following 12 items:

Item No.	Number of Defects
1	8
2	9
3	7
4	8
5	6
6	5
7	8
8	6
9	4
10	3
11	2
12	7

Construct the C-chart.

9. Ten samples have been inspected with sample size = 200. Construct the P-chart for the given data:

Sample No.	Number of Defectives Out of 200 Items Inspected
1	21
2	18
3	14
4	17
5	8
6	9
7	10
8	16
9	20
10	13

10. Differentiate between production planning and production control.
11. Discuss the following:
    (a) Routing
    (b) Scheduling
12. Explain the following in the context of break-even analysis:
    (i) Fixed cost
    (ii) Variable cost
    (iii) Total cost

Derive the break-even quantity, and the quantity to be produced and sold to earn certain specified profit.

13. What do you understand by demand forecasting? Discuss various applications of demand forecasts.

14. Differentiate between qualitative and quantitative methods of demand forecasting.

15. In order to implement moving average method, it is decided to consider six-period averages. The following data show the actual demand in respective periods:

Periods	Actual Demand
1	6
2	4
3	5
4	7
5	4
6	7
7	8
8	6
9	5
10	9
11	7
12	8

Make forecasts for period 7 onward and compare with the actual demands occurred.

16. In the previous exercise, consider the following weightage:

Most recent .... oldest data

25%, 22%, 20%, 15%, 10%, 8%

Apply weighted moving average method.

17. What is the advantage of exponential smoothing? Discuss the role of smoothing constant in this method.

The following data are given:

Current demand = 65

Previous forecast = 58

Smoothing constant = 0.2

Obtain the forecast for the next period.

18. A spark plug manufacturing company wants to correlate its sales with the number of two-wheelers arrived in the market. The company has the data for the past 12 periods as follows:

Periods	Number of Two-Wheelers ($\times 10^5$)	Number of Spark Plugs Sold ($\times 10^5$)
1	5	2
2	9	4
3	6	2
4	12	6
5	10	4
6	9	3
7	11	5
8	13	4
9	14	6
10	13	5
11	9	3
12	8	3

(a) Obtain the relationship between two variables.
(b) In the next period, if the number of two-wheelers to arrive in the market are 15×10^5, estimate the sales of spark plugs.

19. Demand of a product is growing in the following way:

Year	Demand in Numbers ($\times 1000$)
2014	9
2015	11
2016	12
2017	15
2018	16
2019	19
2020	20

It may be convenient to do scaling of the year as follows:

Year – 2017	Demand in Numbers ($\times 1000$)
−3	9
−2	11
−1	12
0	15
1	16
2	19
3	20

Forecast the demand of the product in 2021.

20. Demand of an item is growing as follows:

Year	Demand in Numbers ($\times 10^4$)
2013	12
2014	14
2015	16
2016	19
2017	21
2018	22
2019	25
2020	26

Forecast the demand of the product in the year 2021.

21. Correlation coefficient (r) between the two variables x and y is given by

$$r = \frac{\text{Covariance}}{\sqrt{\text{Variance}(x)}\sqrt{\text{Variance}(y)}} = \frac{S_{xy}}{S_x S_y}$$

where variance (x),

$$S_x^2 = \frac{\sum (x - \bar{x})^2}{n} \quad \text{and} \quad S_x = \sqrt{\frac{\sum (x - \bar{x})^2}{n}}$$

where $\bar{x}$ is the average of data x.
Similarly, variance (y),

$$S_y^2 = \frac{\sum (y - \bar{y})^2}{n} \quad \text{and} \quad S_y = \sqrt{\frac{\sum (y - \bar{y})^2}{n}}$$

where $\bar{y}$ is the average of data y.

$$\text{Covariance, } S_{xy} = \frac{\sum (x - \bar{x})(y - \bar{y})}{n}$$

$$\text{And } r = \frac{\sum (x - \bar{x})(y - \bar{y})}{\sqrt{\sum (x - \bar{x})^2}\sqrt{\sum (y - \bar{y})^2}}$$

A tire manufacturing company wishes to correlate its sales of tires with total number of cars produced and marketed in the country. In the past 10 years, data are available for each year which are as follows:

Year No.	Number of Cars ($\times 10^6$)	Number of Tires Sold ($\times 10^6$)
1	4	1
2	8	3
3	6	2
4	10	4
5	9	3
6	10	3
7	12	5
8	11	3
9	14	6
10	13	6

Find out if there is any correlation.
22. Suggest historical analogy for the following products:
    (a) Automatic washing machine
    (b) Tape recorder
23. With reference to the product life cycle, what measures may be taken before a decline in the demand of product occurs?
24. Explain the Delphi technique considering suitable example.
25. How demand forecasts are useful in the production planning?

BIBLIOGRAPHY

Chase R.B., Aquilano N.J., and Jacobs F.R., *Production and Operations Management*. TMH, 2000.

Nahmias S. *Production and Operations Analysis*. McGraw-Hill, 2001.

Panneerselvam R. *Production and Operations Management*. PHI, 1999.

Smith S.B. *Computer-Based Production and Inventory Control*. Prentice-Hall, 1989.

Index